ATHEIST'S GOD

The Conscious Universe

Unlocking The Secrets of Spirituality Through Modern Physics

AVINASH MISHRA

ISBN 979-8-89544-876-2

Dedication

To my parents,
whose endless love, wisdom,
and guidance have shaped the person I am today.

To my wife,
whose strength and unwavering belief in me have been
my greatest source of inspiration.

To my daughter, Yuvashree,
whose curiosity and joy remind me every day
of the wonder in learning and discovery.

And to you, the reader,
who seeks to explore the depths of knowledge and uncover
the secrets of the universe.
This journey is for all of us.

Contents

Preface

Hello, and thank you for picking up this book. I'm Avinash Mishra, and I've spent much of my life deeply engaged in both the sciences and the spiritual philosophies that have shaped human thought for centuries. The journey to writing this book began with a simple yet profound question that has always intrigued me: Can science and spirituality, often seen as opposing forces, actually converge to offer a new understanding of God?

For as long as I can remember, I've been fascinated by the mysteries of the universe. My love for physics—particularly Quantum Mechanics and String Theory—has driven me to ask some of the biggest questions about existence. At the same time, my exploration of ancient religious texts has opened my mind to the deep wisdom they contain, even though they are often dismissed as purely mystical or outdated.

Over the years, as I taught subjects like Physics, Chemistry, Mathematics, Philosophy, and Business, I noticed something interesting. My students, much like myself, were searching for more than just academic knowledge—they were searching for meaning. Many of them were grappling with questions about the nature of reality, the existence of God, and the purpose of life. This realization sparked a deeper curiosity in me. What if the answers to these profound questions could be found at the intersection of science and spirituality?

This book is the result of six years of rigorous research and contemplation. It wasn't a straightforward journey—there were moments when the complexity of the subject matter felt overwhelming. But each challenge strengthened my resolve to bring these ideas to light in a way that's accessible to everyone. My goal was never to write a purely academic book, nor a purely spiritual one. Instead, I wanted to create something that would bridge the gap between the two, offering readers a new way to think about God—one that is grounded in the logic of modern physics yet resonates with the spiritual truths that have guided humanity for millennia.

As you read through these pages, I hope you'll find not just answers, but also inspiration to explore these concepts further. The universe is far more complex and interconnected than we can ever fully understand, but it is in this complexity that we find the most profound truths. This book is an invitation to look at the universe, and at God, through a new lens—one where science and spirituality are not in conflict but in harmony.

Thank you for joining me on this journey. I hope it challenges your thinking, deepens your understanding, and opens your mind to the endless possibilities of what we might discover when we allow these two worlds to come together.

Prologue

In the vast expanse of the universe, countless mysteries lie hidden, waiting to be unraveled. For centuries, humanity has looked to the skies, the earth, and within themselves, seeking answers to the profound questions that define our existence. Who are we? What is our purpose? Is there a higher power guiding the cosmos, or are we simply the product of chance in an indifferent universe?

These questions have given rise to countless religious doctrines, philosophical musings, and scientific theories. Yet, despite the advancements in our understanding of the physical world, the nature of God remains as elusive as ever. The ancient texts speak of a divine presence that shapes and sustains all things, while modern physics presents us with a reality that is stranger and more complex than anything our ancestors could have imagined.

This book is born from the intersection of these two realms—the spiritual and the scientific. It is a journey that began with a simple question: Can the principles of modern physics, particularly Quantum Mechanics and String Theory, offer a new understanding of God? Can science, which often seems at odds with religion, actually bring us closer to the divine?

In the chapters that follow, we will explore these questions with a spirit of open-minded inquiry. We will delve into the depths of physics, examining the latest theories that describe the fabric of reality, and we will revisit the ancient wisdom of religious traditions, searching for the common threads that unite them with modern science.

This book does not seek to diminish the spiritual experiences that have shaped human history, nor does it aim to undermine the scientific discoveries that have expanded our knowledge of the universe. Instead, it offers a new perspective—one where science and spirituality are not opposing forces but complementary paths leading to a deeper understanding of the divine.

As you turn the pages, I invite you to set aside preconceived notions and join me on a journey of discovery. Together, we will explore the possibility that the universe is not a random, lifeless expanse but a living, conscious entity—a manifestation of what we might call the "Atheist's God." In doing so, we may find that the mysteries of existence are not as separate as they seem, but are instead part of a single, coherent whole.

Acknowledgements

I would like to express my deepest gratitude to everyone who supported and guided me throughout the journey of writing this book.

First and foremost, I want to thank my family for their unwavering support and patience, especially during the many late nights and early mornings spent on this project.

I am also immensely grateful to my students at Noetic Education Pvt Ltd and StudyAtom, whose insights and discussions played a crucial role in shaping the ideas presented in this book.

A special thank you goes to my publisher, whose belief in this project and guidance throughout the process have been invaluable.

Lastly, I want to acknowledge the many thinkers, scientists, and spiritual leaders whose works have inspired and challenged me to explore the ideas that form the foundation of this book.

Thank you all for being part of this journey with me.

Introduction

A Journey from Curiosity to Understanding

"Somewhere, something incredible is waiting to be known."

– Carl Sagan

Welcome to an exploration that lies at the intersection of science, consciousness, and the very nature of reality. This book is not just a presentation of facts, theories, or philosophies; it's a guide to understanding the deeper connections that bind the universe together. Whether you come from a scientific background or are simply curious about the mysteries of existence, this journey is for you.

The Purpose of This Book

The purpose of this book is to show that the Universe is a living entity, filled with consciousness and intelligence. Through logical reasoning and scientific insights, this book offers a fresh perspective on understanding the concept of God as presented by every religion. It answers the practical questions that a logical person might have about their faith, providing clarity and deeper meaning.

By connecting the findings of modern physics with spiritual ideas, this book helps bridge the gap between a normal human mind and their true understanding of God. It guides you to connect more deeply with your surroundings, offering solutions to many of life's challenges.

Lastly, this book simplifies the most complex theorems and concepts of Modern Physics, making them accessible to everyone. Whether you are seeking answers, connection, or a better understanding of the universe, this book aims to be your guide.

Structure of the Book

This book is organized into two main modules: **Module A** and **Module B**.

Module A is designed to teach you all the important concepts of Modern Physics, starting from the very basics of science. It's perfect for anyone who has little to no background in this field. This module acts as a guide, providing you with the logical foundation needed to understand the teachings in Module B.

Module B builds on the logic and knowledge from Module A, using it to address questions related to religious beliefs, the concept of God, and spirituality. This module is not just about understanding; it also serves as a practical guide to solving life's problems in a logical and meaningful way. With the understanding from Module A, you'll be equipped to navigate the insights and solutions offered in Module B. Additionally, we will explore the compelling idea that the universe itself is a living entity, further deepening our understanding of the connection between science and spirituality.

Who Should Read This Book?

Whether you are a student, a professional, a lifelong learner, or someone simply fascinated by the mysteries of existence, this book is crafted for you. No prior knowledge of physics, philosophy, or mathematics is required. All you need is curiosity and an open mind.

What You Will Discover

1. **The Basics of Science:** We start by laying a solid foundation, exploring the scientific method, the role of mathematics, and the essential concepts of space, time, matter, and energy.

2. **Quantum Mechanics and Beyond:** Dive into the mind-bending world of quantum mechanics, where reality is far stranger than fiction. Discover how these principles challenge our understanding of the universe and our place within it.

3. **String Theory and Unified Theories:** Journey into the heart of modern physics with an exploration of string theory and its promise of a unified theory of everything.

4. **Consciousness and the Universe:** Explore the fascinating connections between quantum mechanics, consciousness, and the broader concept of super-consciousness, where science and spirituality converge.

5. **Practical Applications:** Learn how the concepts discussed in this book can be applied to everyday life, helping you overcome challenges, develop new skills, and achieve a deeper understanding of yourself and the world around you.

6. **Scientific Aspects of God:** Using multiple theories of modern physics, we delve into the scientific aspects of the real concept of God, offering a fresh perspective that bridges science and spirituality.

A New Perspective

This book isn't just about learning scientific concepts—it's about seeing the world in a new light. By the end of this journey, you'll have a deeper appreciation for the universe, a better understanding of the role of consciousness in shaping reality, and perhaps most importantly, a realization that science and spirituality are not opposing forces but different ways of understanding the same fundamental truth.

As you turn these pages, remember that every discovery, every insight, and every question is a step toward a greater understanding of the incredible universe we are all a part of. So, let's embark on this journey together, with curiosity as our guide and the search for truth as our destination.

Welcome to the adventure.

Module A

The Basics of Science

CHAPTER 1

Foundations of Science and the Universe

Introduction

"Science is a way of thinking much more than it is a body of knowledge."

– Carl Sagan

Welcome to Module A. Before we dive in, I want to let you know that this entire module is all about science—no philosophical ideas here. But don't worry, I've made sure to explain everything in the simplest way possible, so even if you're not from a science background, you'll be able to follow along easily.

There might be moments when the scientific concepts seem a bit dry, but stick with it—I promise your curiosity will grow with each chapter. And trust me, Module B is going to be worth the wait! However, to truly appreciate what's coming in Module B, it's important to understand the basics covered in Module A. So, don't skip ahead—this module is key to grasping the entire book and its core ideas.

Before diving into the specific topics of quantum mechanics, string theory, and other advanced scientific concepts, it's essential to lay a solid foundation. This chapter will explore the basic principles that underlie all of science, providing a framework that will help you understand the more complex ideas we'll discuss later.

What is Science?

At its core, science is the study of the natural world through observation, experimentation, and analysis. It's a systematic way of asking questions, gathering data, and drawing conclusions. Science helps us understand how the universe works, from the smallest particles to the largest galaxies.

1. The Scientific Method: A Step-by-Step Process

The scientific method is the foundation of all scientific inquiry. It's a process that helps scientists test ideas and theories in a structured way.

- **Observation:** The first step in the scientific method is observation. Scientists observe the world around them, identifying patterns or phenomena that need explanation. For example, you might notice that the sky is blue during the day and wonder why that is.

- **Question:** Once an observation is made, the next step is to ask a question. In our example, the question might be, "Why is the sky blue?"

- **Hypothesis:** A hypothesis is an educated guess that attempts to answer the question. It's a proposed explanation that can be tested. For example, a hypothesis could be, "The sky appears blue because of the way sunlight interacts with the Earth's atmosphere."

- **Experiment:** To test the hypothesis, scientists conduct experiments. These experiments are designed to collect data and provide evidence that either supports or refutes the hypothesis.

- **Analysis:** After gathering data, scientists analyze the results. They look for patterns or correlations that either confirm or disprove the hypothesis.

- **Conclusion:** Based on the analysis, scientists draw a conclusion. If the data supports the hypothesis, it becomes a theory. If not, the hypothesis is revised, and the process begins again.

Example:

Imagine you're baking a cake, and you want to find out what happens if you add more sugar than the recipe suggests. You observe that the cake turns out sweeter and maybe even fluffier. You hypothesize that sugar affects both the sweetness and the texture. You experiment by baking several cakes with varying amounts of sugar, then analyze the results to see how each cake turns out. Finally, you conclude that sugar not only makes the cake sweeter but also changes its texture.

2. Laws and Theories: The Building Blocks of Science

In science, laws and theories are fundamental concepts that help us understand the natural world.

- **Scientific Laws:** Scientific laws describe what happens under certain conditions but do not explain why it happens. They are statements based on repeated experiments and observations. For example, Newton's Law of Gravity describes the force of attraction between two masses but doesn't explain why gravity exists.

- **Scientific Theories:** Theories, on the other hand, explain why phenomena occur. They are well-substantiated explanations of some aspect of the natural world, based on a body of evidence. For example, the Theory of Evolution explains how species change over time through natural selection.

The Language of Science: Mathematics

Mathematics is often called the "language of science" because it provides the tools needed to describe and understand the universe. Whether you're calculating the speed of a falling object or understanding the orbit of a planet, math is essential.

3. Numbers and Units

Numbers are the basic building blocks of mathematics. In science, numbers are used to measure and quantify the world around us. These measurements are always accompanied by units, which tell us what exactly is being measured (e.g., meters for length, seconds for time, kilograms for mass).

Example:

If you say something is 5 meters long, **the number** (5) tells you the quantity, and **the unit** (meters) tells you what is being measured.

4. Equations and Formulas

Equations are mathematical statements that show the relationship between different quantities. Formulas are specific types of equations that are used to solve particular problems.

Example:

The equation F = ma is a famous formula from physics that states that force (F) is equal to mass (m) multiplied by acceleration (a). This formula allows us to calculate the force acting on an object if we know its mass and acceleration. (Acceleration is the rate at which an object's speed changes over time. If something is speeding up, slowing down, or changing direction, it's experiencing acceleration.)

The Nature of Reality: Space, Time, and Matter

To understand our universe, we need to grasp three fundamental concepts: space, time, and matter. These are the building blocks of everything in the physical world.

1. Space: The Framework of the Universe

Space is the vast, seemingly endless expanse in which all objects and events exist. It's not just empty; space can be bent or curved by the presence of mass and energy. This idea comes from Einstein's Theory of General Relativity, which fundamentally changed how we understand space and gravity.

Einstein's Theory of General Relativity tells us that space is not just a static stage where things happen; it's dynamic and flexible. When a massive object, like the Earth or the Sun, is placed in space, it causes the space around it to curve or warp. This curvature of space is what we perceive as gravity. The more massive the object, the more it bends space, and the stronger the gravitational pull.

Example:

Think of space as a soft, stretched-out sheet. When you place a heavy ball on this sheet, it creates a dip or curve around the ball. If you then roll a smaller ball near the heavy one, it will naturally move toward the heavier ball because of the curve in the sheet. Similarly, planets orbit the Sun because the Sun's mass curves the space around it, creating a "path" for the planets to follow.

2. Time: The Flow of Events

Time is the progression from the past, through the present, and into the future. It's often seen as a constant flow, but according to Einstein's Theory of

Relativity, time can actually be stretched or compressed depending on how fast you are moving or how strong the gravity is around you.

Einstein's Theory of Relativity shows that time is not the same for everyone. If you move at a very high speed or are in a place with strong gravity, time will pass differently for you compared to someone who is moving slower or is in a place with weaker gravity. This phenomenon is known as time dilation.

Example:

Imagine you are traveling in a spaceship at nearly the speed of light. For you, time would pass more slowly than for someone who is on Earth. If you travel for what feels like one year in your spaceship, you might return to find that many years have passed on Earth. This effect, while mostly noticeable at very high speeds or in extreme gravitational fields, shows that time is not as fixed as we might think.

3. Matter: The Substance of the Universe

Matter is anything that has mass and occupies space. It's the "stuff" that makes up everything in the universe, from the smallest particles to the largest stars. Matter is made of atoms, which are themselves composed of even smaller particles like protons, neutrons, and electrons.

At the core of all matter are atoms, which are the basic units of chemical elements. Each atom has a nucleus, made up of protons and neutrons, surrounded by electrons that orbit the nucleus. These atoms come together to form the various substances we see around us. The properties of matter—such as solidity, liquidness, or gaseousness—depend on how these atoms are arranged and how they interact with each other.

Example:

Consider a wooden chair. It seems solid and unchanging, but it's made up of countless atoms that are in constant motion. The chair's mass and its ability to support weight come from the collective mass of all these atoms, which are held together by atomic bonds. Even though we can't see these atoms with the naked eye, they are the reason the chair has the properties that it does.

Space, time, and matter are the three essential components of the universe. Space provides the structure in which everything exists, time allows for the sequence of events, and matter is the substance that makes up everything. Understanding these concepts helps us comprehend the nature of reality and the fundamental workings of the universe.

Energy: The Driving Force

Energy is the ability to do work, and it comes in many forms—such as kinetic energy (the energy of motion), potential energy (stored energy), thermal energy (heat), and electromagnetic energy (light).

1. Conservation of Energy

One of the most important principles in physics is the conservation of energy, which states that energy cannot be created or destroyed, only transformed from one form to another.

Example:

When you lift a book, you're giving it potential energy. If you let go, that potential energy is converted into kinetic energy as the book falls. When it hits the ground, some of that energy is transferred into sound and heat.

2. The Relationship Between Matter and Energy

Matter and energy are closely related. In fact, Einstein's famous equation $E=mc^2$ shows that matter can be converted into energy and vice versa. This equation explains how stars shine—they convert mass into energy through nuclear fusion.

Example:

In a nuclear reaction, a small amount of matter is converted into a large amount of energy, which is why nuclear power plants can produce so much energy from relatively small amounts of fuel.

Conclusion: The Building Blocks of Knowledge

This chapter has laid the groundwork for understanding the universe from a scientific perspective. We've explored what science is, how the scientific method works, and the fundamental concepts of space, time, matter, and energy. Understanding these basics is essential as we move forward into more complex topics like quantum mechanics and string theory.

As you continue through the chapters of Module A, remember that science is not just a collection of facts—it's a way of thinking and exploring the world. The tools and concepts we've discussed here will help you navigate the more advanced ideas that follow, providing a strong foundation for understanding the incredible complexity and beauty of the universe.

CHAPTER 2

Quantum Mechanics and Quantum Entanglement

Introduction

In the early 20th century, scientists discovered a new way of understanding the world that didn't follow the usual rules of classical physics. This mysterious and strange realm is called the quantum world. It's a place where particles can be in multiple places at once, where things can affect each other instantly across vast distances, and where our common understanding of how the world works seems to fall apart.

In this chapter, we'll explore the basics of quantum mechanics and one of its most fascinating features—quantum entanglement. This will help us understand how the smallest parts of our universe behave and why this is so different from what we experience in our everyday lives.

The Beginning of Quantum Mechanics

In the early 1900s, scientists noticed that the existing laws of physics, like those developed by Newton and Maxwell, couldn't explain everything they observed when they looked at very small things like atoms. This led to the creation of quantum mechanics, a new way of thinking about how nature works on the smallest scales.

Key Ideas in Quantum Mechanics

Quantum mechanics is the branch of physics that studies the behavior of the tiniest particles in the universe—particles like electrons, protons, and even smaller ones like quarks and photons. Quarks, for example, are the building blocks that make up protons and neutrons, while photons are particles of light.

In the quantum world, these tiny particles behave in ways that are very different from what we see in our everyday lives. For instance, a particle can exist in two places at once, or it can instantly affect another particle far away.

These strange behaviors are what quantum mechanics seeks to understand and explain.

Quantum mechanics is essential for understanding how these extremely small particles interact with each other and the forces that govern them. This understanding helps explain many fundamental aspects of the universe, from the structure of atoms to the nature of light and energy.

Wave-Particle Duality

Imagine if something could be two things at once. That's the surprising truth about electrons, tiny bits of matter that are inside everything around us. In the world of quantum mechanics, electrons can act like small, solid objects (like marbles) and also like waves (like ripples on a pond). This might sound strange, but let's break it down.

Electrons as Little Marbles:

Picture a marble rolling across a table. You can see it, touch it, and even flick it to make it move. Electrons can behave in a similar way. For instance, if you shoot a bunch of electrons at a metal surface, they'll hit it like tiny marbles, causing other electrons to move. This shows that electrons can behave like small, solid particles.

Electrons as Ripples in Water:

Now, think about dropping a stone into a pond. The stone creates ripples that spread out across the water. These ripples show how waves move, spreading and overlapping. Amazingly, electrons can behave like this too! When electrons pass through two tiny slits side by side, they don't just travel straight through like marbles. Instead, they spread out and create a pattern, similar to the ripples in the pond. This wave-like behavior is one of the most surprising things about quantum mechanics.

How is This Possible?

Think of it like this: Imagine you're at a sports event, and the crowd does a wave. From far away, it looks like a smooth wave moving through the stands.

But if you look closely, you'll see that the wave is actually made up of individual people standing up and sitting down.

Now, let's connect this to particles like electrons and photons. When we look at them one way, they appear as tiny particles, just like seeing each person in the crowd. But when we look at them in another way—like how they move or how they spread out—they behave like waves, just like the wave moving through the crowd.

Both ways of seeing them are true, depending on how we observe them. It's not magic; it's just a different way of looking at the same thing. This is what scientists mean when they say that particles like electrons and photons can act as both particles and waves. It's a fundamental part of how nature works on the smallest scales.

Quantization of Energy: Understanding Energy in Chunks

In the world of quantum mechanics, energy doesn't come in any random amount—it comes in specific, fixed chunks. Let's think of energy like money. Just as you can only have whole coins or notes, and not half a coin, energy comes in certain fixed amounts too.

Imagine an electron inside an atom like someone standing on a ladder. Each step on the ladder represents a specific amount of energy. The electron can jump up to a higher step or down to a lower one, but it can't stop in between the steps. It can only move by a whole step at a time, just like you can only move between whole coins and not half coins.

So, quantization of energy is like saying energy comes in "packets" or "chunks" that can't be divided into smaller parts, just like you can't split a coin into half and still use it. This is a fundamental idea in quantum mechanics, helping explain how particles behave at the smallest scales.

The Uncertainty Principle

The Uncertainty Principle is a concept in quantum mechanics that says we can't know everything about a tiny particle, like an electron, at the same time. Specifically, if we try to measure where the particle is (its position), we can't know exactly how fast it's moving (its momentum), and vice versa. The more precisely we know one of these things, the less precisely we can know the other.

Imagine you're trying to take a clear photo of a speeding car at night. If the car is moving very fast, the photo might come out blurry, so it's hard to tell exactly where the car is in the picture. But if the car is parked, you can take a sharp, clear picture, knowing exactly where it is—but now, you don't have any information about how fast it's moving because it's not moving at all.

In the quantum world, this isn't just a matter of better tools or techniques; it's a fundamental limit of nature. No matter how good our measurements get, we can never know both the exact position and exact momentum of a particle at the same time. This principle is one of the key ideas in quantum mechanics, showing that at the smallest scales, there's a built-in uncertainty in how we can describe the world.

The Wave Function and Schrödinger's Equation

The Wave Function: In quantum mechanics, the wave function is a mathematical tool used to describe the state of a particle, like an electron. It tells us everything we can know about the particle's position, momentum, and other properties. But instead of giving a definite answer, the wave function gives us probabilities—essentially telling us where the particle is likely to be found and what it's likely to be doing.

Imagine the wave function as a map that shows the chances of finding a particle in different places. For example, in some areas, the map might show a high probability (like a bright spot), meaning the particle is likely to be there. In other areas, the probability might be low (like a dim spot), meaning the particle is less likely to be found there.

Schrödinger's Equation: Schrödinger's Equation is the fundamental equation in quantum mechanics that tells us how the wave function changes over time. It's like a rule book that governs the behavior of the wave function.

Think of Schrödinger's Equation as a recipe that takes the current state of a particle (described by the wave function) and predicts how that state will evolve in the future. It helps us understand how particles move, interact, and change within the quantum world.

In simpler terms, if you know the wave function of a particle at a certain time, Schrödinger's Equation allows you to calculate how that particle's wave function—and therefore its behavior—will change as time goes on. This equation is key to understanding and predicting the behavior of particles at the quantum level.

Superposition: Understanding the Quantum World

In quantum mechanics, there's a concept called superposition that's different from what we see in our everyday world. It's the idea that a tiny particle, like an electron, can be in multiple states at the same time.

Think of Different Choices

Imagine you're deciding between two options, like sitting in two different chairs. In the regular world, you can only sit in one chair at a time. But in the quantum world, an electron can be like it's sitting in both chairs at once. It's not that the electron is moving back and forth really fast; it's actually in both places at the same time.

What Happens When We Look

However, this strange situation changes the moment we try to check or measure the electron. As soon as we look, the electron "chooses" one chair, and that's the only place we'll find it. The other possibility, the other chair, just disappears.

The idea that a particle can be in two states at the same time, known as superposition, is indeed difficult to understand because it's so different from how things work in our everyday world.

Here's a way to think about it: In the world of quantum mechanics, particles like electrons don't behave the same way as larger objects, like a chair or a ball, which can only be in one place or state at a time. Instead, these tiny particles exist in a sort of "cloud" of possibilities. Before we observe or measure them, they don't settle into one definite state. Instead, they sort of "spread out" over all the possible states they could be in.

It's as if the particle is considering all its options at once. It's not that the particle is physically in two places or states at the same time, like splitting itself in half. Instead, it's more about probabilities: the particle has a certain chance of being in each possible state, and until we measure it, it stays in this fuzzy, uncertain condition.

Once we measure the particle, the superposition collapses, and the particle "chooses" one of the possible states. This is not something we see in the larger world, which is why it seems so strange. But it's a fundamental part of how the very small things in the universe work, according to quantum mechanics.

Double-Slit Experiment: The Curious Behavior of Particles

So, while it may seem impossible or counterintuitive, it's a real effect that can be understood through a famous experiment called the 'Double-Slit Experiment.' Let's study it to see how this works.

The double-slit experiment is a famous example that helps us understand the strange behavior of tiny particles like electrons. Here's how it works:

The Setup: Imagine a wall with two small openings, called slits, and you're shooting tiny particles, like electrons, at the wall. Behind the wall is a screen that shows where the electrons land after passing through the slits.

What Happens Without Watching

When you shoot electrons at the wall, you might expect to see two spots on the screen where the electrons passed through the slits. But something surprising happens instead. The screen shows a pattern of light and dark bands, as if the electrons are spreading out and overlapping after passing through the slits. This pattern looks like something you'd see if waves of water were passing through the slits and then overlapping, creating a series of bands on the screen. It's as if each electron is behaving like a wave, passing through both slits at once and spreading out as it moves.

What Happens When You Watch:

Now, if you place a detector near the slits to see which slit each electron goes through, something changes. Instead of creating the pattern of bands, the electrons start to behave like little balls again, making just two spots on the screen—one behind each slit. The moment you observe them, the electrons act like solid particles, choosing to go through one slit or the other.

Observation as Interaction

In the quantum world, observation isn't just about looking at something from a distance. To observe a quantum particle like an electron, you have to interact with it in some way. This interaction usually happens through a medium, like light or other particles, that "touches" the electron so that we can detect its position or state.

The Role of the Observer and the Medium

1. **The Observer:** The observer is the entity (like a scientist or a measuring device) trying to gain information about the quantum particle.

2. **The Medium of Observation:** This could be light, another particle, or any tool used to probe the quantum system. For example, if you shine light on an electron to observe it, the photons (light particles) interact with the electron.

Why This Interaction Changes the Particle

Before observation, the electron exists in a superposition—a cloud of all possible states it could be in. But when you try to observe it, the medium of observation (like light) interacts with the electron. This interaction disturbs the electron, causing it to "choose" a specific state.

Think of it like trying to catch a soap bubble: the moment you touch it, you change it or even pop it. Similarly, the act of observing the electron—by interacting with it—collapses the superposition of possibilities into one definite outcome.

The Collapse of Possibilities

The "collapse" happens because the interaction disturbs the delicate balance of possibilities the particle was in. The electron, once free to exist in multiple states, is forced into one specific state due to this interaction. This is why, after observation, the electron behaves like a particle and not like a wave.

In short, what's special about observation in quantum mechanics is that it forces the system to leave the strange quantum realm of multiple possibilities and enter the classical world where we see only one reality at a time.

Quantum Entanglement

Quantum entanglement is a phenomenon in quantum mechanics where two or more particles are created or interact in such a way that their properties become linked. Once entangled, the state of one particle is automatically connected to the state of the other, no matter how far apart they are. This link means that if you measure one particle, you immediately know the state of the other, even if it's on the opposite side of the universe. This profound connection suggests that the universe is not a collection of isolated objects but a web of interconnected entities, all tied together by this invisible thread of entanglement.

The Experiment: Entangled Photons

One famous real-world example of quantum entanglement is an experiment involving pairs of entangled photons (particles of light). Here's how it works:

- Scientists can create a pair of entangled photons by passing a beam of light through a special crystal. This process splits one photon into two entangled photons. These two photons now have linked properties, such as their polarization (the direction in which they oscillate).

- Once the photons are entangled, they can be sent in opposite directions, far away from each other—sometimes even miles apart.

- If scientists measure the polarization of one photon, they instantly know the polarization of the other photon, even if it's far away. For instance, if the first photon is measured and found to have a vertical polarization, the second photon will instantly have a horizontal polarization, regardless of the distance between them.

The Key Point

This result happens faster than any signal could travel between the two photons, meaning their states are connected in a way that defies classical physics. It's as if the two photons are communicating instantly, even though they're far apart, demonstrating the reality of quantum entanglement.

This experiment shows that entangled particles are connected in a way that transcends the normal limits of space and time, supporting the idea that quantum entanglement is a fundamental feature of the universe.

EPR Paradox

The idea of entanglement was famously highlighted by Albert Einstein, Boris Podolsky, and Nathan Rosen in 1935, in what is now known as the EPR paradox. They argued that quantum mechanics might be incomplete because it couldn't fully explain how entanglement works—what Einstein dubbed "spooky action at a distance." They believed there must be hidden variables that determine the states of entangled particles, rather than the particles being mysteriously linked across distances.

Bell's Theorem

In the 1960s, physicist John Bell developed a way to test whether these hidden variables actually existed. His work, known as Bell's theorem, demonstrated that no local hidden variable theories could fully explain the predictions made by quantum mechanics. When scientists conducted experiments based on Bell's theorem, the results consistently supported quantum mechanics and ruled out the idea of local hidden variables. This confirmed that entanglement is real and suggests that the universe operates as a unified, interconnected whole, where everything is linked by a deeper, unseen reality.

How Entanglement Works

Entanglement begins when particles become connected in a special way. Here's how this can happen:

- **When Particles Interact and Then Separate:** Imagine two particles meeting and interacting, much like two friends shaking hands. After their interaction, they become entangled, meaning they remain connected even as they move apart.

- **When a Particle Splits into Two:** Another way entanglement happens is when one particle splits into two or more particles. For example, think of a photon—a particle of light. If this photon splits into an electron and a positron, these new particles can become entangled, meaning that whatever happens to one of them instantly affects the other, no matter the distance between them.

In simple terms, entanglement creates a special link between particles, binding them together in such a way that a change in one instantly causes a change in the other. This connection is a key piece of the puzzle in understanding how the universe operates as a living, interconnected system.

Entangled States

When particles are entangled, they become so closely linked that they cannot be considered separately. Instead, they must be seen as parts of a single system, described by a "wave function."

The wave function is like a blueprint that encompasses all the entangled particles together. It shows the different possibilities for what might happen when we observe the particles. But here's the crucial point: this blueprint can't be divided into separate pieces for each particle. The particles are so interconnected that their behaviors are tied together in one unified picture.

So, instead of each particle having its own independent description, they share a combined description that reflects their connection. This is a glimpse into the deeper reality where everything is connected, aligning with the idea of a living, conscious universe.

Measurement and Collapse

When two entangled particles are observed, they are connected in such a way that whatever happens to one instantly affects the other. This connection is captured by the wave function.

When we measure one of these entangled particles, something remarkable happens: the wave function collapses. This means that the uncertainty about what the particles are doing disappears, and their states become definite.

Example:

Imagine two entangled particles with spins (a property of particles that can be thought of as a tiny magnetic direction). If these particles are entangled so that their spins are always opposite, measuring the spin of one particle immediately determines the spin of the other. If you measure the first particle's spin as "up," the wave function collapses, and the other particle's spin is instantly "down," no matter how far apart the particles are. This immediate connection is a reflection of the universe's underlying unity.

Experiments Demonstrating Entanglement

Several key experiments have confirmed the reality of quantum entanglement:

- **Aspect Experiment (1982):** In 1982, Alain Aspect and his team conducted an experiment to test the reality of quantum entanglement. They used entangled photons—particles of light—to see if what happened to one photon would instantly affect the other, even across vast distances.

  The results were clear: the photons behaved exactly as quantum mechanics predicted, with no evidence of hidden variables controlling them. This experiment provided strong proof that entanglement is real and that particles are indeed connected across distances, supporting the idea that the universe is interconnected at its most fundamental level.

- **Teleportation Experiments:** Quantum teleportation involves transferring the state or information of one particle to another, even if they are far apart, using quantum entanglement.

  In 1997, scientists at the University of Innsbruck successfully "teleported" the state of a photon to another photon several kilometers away. They didn't move the photon itself but transferred its information instantly. This experiment demonstrated that quantum entanglement could have real-world applications, like transmitting information across distances, further supporting the concept of an interconnected universe.

- **Loophole-Free Bell Tests:** In recent years, scientists have worked to eliminate any doubts in experiments testing quantum entanglement, particularly concerning Bell's theorem. In 2015, researchers at Delft University of Technology conducted experiments designed to close any "loopholes" that could question the results.

  The findings confirmed quantum entanglement with incredible precision, providing the strongest evidence yet that entangled particles behave as predicted. These results leave no room for doubt, further solidifying the idea that the universe is a deeply interconnected system.

Philosophical Implications of Quantum Entanglement

Quantum entanglement also raises profound philosophical questions about reality, interconnectedness, and the role of information in the universe:

- **Reality and Measurement:** In quantum mechanics, entanglement suggests that reality at the quantum level doesn't have definite properties until it is observed. This idea challenges our traditional view of reality, proposing that observation—and thus consciousness—plays a crucial role in determining what exists.

  Example: Imagine a pair of socks, one black and one white, in different drawers. In quantum mechanics, it's as if the socks don't decide their color until you open a drawer. Before you look, the sock exists in a state of both black and white! This concept challenges our understanding of reality, suggesting that consciousness itself might be what brings specific properties into existence.

- **Everything is Connected (Holism):** Quantum entanglement suggests that everything in the universe is connected in some fundamental way. This aligns with the idea of Super-consciousness—the living force that connects all strings in string theory, holding the universe together as a unified whole.

  Example: Think of two friends who are so in sync that they can finish each other's sentences, even when they're far apart. In quantum mechanics, this is like two particles being "in sync" with each other, no matter how far away they are. This idea aligns with certain spiritual and philosophical beliefs that the universe is interconnected, with every part influencing the whole.

- **The Role of Information:** In quantum mechanics, "information" plays a crucial role. Some scientists believe that "information" might be as fundamental to the universe as matter and energy, aligning with the idea that the universe itself is a living, conscious entity.

Example 1: Think of a secret code or message in your mind. The message isn't a physical thing you can touch, but it's still real because it conveys information. In quantum mechanics, particles seem to communicate with each other instantly. When particles are entangled, the "information" about one particle is shared with the other instantly, no matter how far apart they are. This reflects the idea that the universe is a web of interconnected "information" and consciousness.

Example 2: Consider a video game where the entire game world is created by code. This code is the "information" that tells the computer what to display on the screen. Without this code, the game wouldn't exist. Some scientists propose that the universe operates similarly, where "information" is a fundamental part of reality. In quantum mechanics, when particles are entangled, they share "information" instantly, suggesting that "information" is a key element in the fabric of reality itself.

Conclusion

Quantum mechanics and quantum entanglement reveal a universe that operates on principles very different from the ones we know in our everyday lives. At the smallest scales, particles behave in ways that are probabilistic, interconnected, and often counterintuitive.

Understanding these principles is crucial because they are the foundation of much of modern science and technology. Quantum mechanics isn't just a theory; it's a well-tested framework that has led to the development of everything from semiconductors in our computers to the future potential of quantum computers.

As we continue to study quantum mechanics, we will explore how these principles might connect to broader ideas in physics and beyond. For now, it's important to grasp the basics of how the quantum world operates. This will set the stage for deeper explorations in the chapters that follow.

Note: Quantum mechanics rules and effects are mostly observable at the microscopic scale, where particles can exist in multiple states simultaneously (superposition) and behave unpredictably. However, in our everyday life, we deal with large objects made up of countless particles, where these quantum effects average out and become negligible due to interactions with the environment, a process known as decoherence. This is why we don't notice quantum phenomena like superposition or entanglement in the macroscopic world.

CHAPTER 3

Quantum Field Theory

Introduction

Quantum Field Theory (QFT) might sound complicated, but it's a powerful tool that helps us understand the tiniest parts of the universe and how they interact. Think of QFT as a bridge connecting the strange world of quantum mechanics with the bigger picture of how everything in the universe fits together. In this chapter, we'll break down the basics of QFT, explore why it's important, and see how it helps us understand the universe on a deeper level.

Traditional View: The Universe Made of Particles

Common Understanding: We usually learn that the universe is made up of particles like electrons, protons, and neutrons. These particles are thought of as tiny, solid objects that move through space and interact with each other.

Particles as Building Blocks: In this view, particles are the fundamental building blocks of everything. For example, atoms are made of particles, and molecules are made of atoms.

Quantum Field Theory (QFT) Perspective: The Universe Made of Fields

Fields as the Foundation: In QFT, the universe is made up of fields rather than just particles. A field is something that exists everywhere in space and can have different values at different points. Imagine a field like an invisible ocean that fills all of space, with ripples and waves moving through it.

Particles as Ripples in Fields: Instead of seeing particles as tiny objects, QFT describes them as excitations or "ripples" in their respective fields. For example, an electron isn't a tiny object moving through space; it's a ripple in the "electron field" that fills the universe.

Analogy: Imagine a calm ocean. If you drop a pebble into it, waves spread out from where the pebble hits the water. In QFT, these waves are like particles. The ocean itself is the field, and the waves are the particles we observe.

Example: The electromagnetic field is everywhere, and when it's excited, we observe photons (particles of light). So, a photon is just a ripple in the electromagnetic field.

Why This Matters

Unified Understanding: Thinking of the universe as being made up of fields helps us understand how different forces and particles are connected. For example, the same QFT principles apply to the electromagnetic field (which produces light) and the electron field (which produces electrons).

Dynamic and Interconnected: Fields can interact with each other, and these interactions produce the phenomena we observe, like forces between particles or the creation and destruction of particles. This view shows that the universe is dynamic and interconnected at the most fundamental level.

Creating and Destroying Particles in QFT

In QFT, the universe is like a constantly changing sea of fields where particles are not fixed objects but temporary excitations or disturbances in these fields. Just as bubbles can form and pop in boiling water, particles can appear and disappear within these fields. This is a fundamental aspect of how the quantum world operates.

1. Creating Particles: The Role of Creation Operators

How It Works: Think of a still body of water. Now, imagine blowing air into the water to create a bubble. The bubble didn't exist before; it was created by your action. Similarly, in the quantum world, the creation operator acts on a quantum field to "blow up" a particle, making it appear out of the field's energy.

Why It's Important: Creation operators are essential for understanding how particles are born in the quantum world. They explain how particles can emerge in high-energy environments like those found near black holes or during particle collisions in accelerators.

2. Destroying Particles: The Role of Annihilation Operators

How It Works: Just as easily as you can create a bubble, you can also pop it, making it disappear back into the water. In QFT, the annihilation operator does something similar for particles. It causes the particle to vanish and return to the quantum field.

Why It's Important: Annihilation operators help us understand how particles can disappear from the observable universe. This process is important for phenomena like particle-antiparticle annihilation, where particles and their opposites destroy each other, releasing energy.

How Fields Move and Change in QFT

To understand how fields and particles behave, scientists use two main approaches: the Lagrangian formulation and the Hamiltonian formulation. These approaches are like two different tools that offer unique ways to explore and predict the behavior of fields in the universe.

The Recipe: Lagrangian Formulation

- **How It Works:** The Lagrangian is like a recipe that tells us how fields mix, interact, and evolve. It's a function that depends on the properties of the fields, such as their position and velocity, and it encapsulates all the information needed to describe the dynamics of a field.

- **Why It's Important:** By using the Lagrangian, scientists can derive the equations of motion for the fields, predicting how they will evolve over time and how particles will emerge from these fields.

The Map: Hamiltonian Formulation

- **How It Works:** The Hamiltonian is like a map that tracks the flow of energy within a field over time. It represents the total energy of the system, including both kinetic and potential energy, and it describes how the energy in a field is distributed and changes as the field evolves.

- **Why It's Important:** The Hamiltonian formulation is used to calculate the future states of a field based on its current energy configuration, helping scientists predict how the universe will unfold from one moment to the next.

The Standard Model: The Universe's Blueprint

Quantum Field Theory is the foundation of the Standard Model, which serves as the universe's blueprint for how all the tiny building blocks (particles) fit together and interact.

Types of Particles

The Standard Model classifies particles into two main types:

- **Fermions:** These are the "building blocks" of matter, like the bricks in a wall. They include particles like quarks (which combine to form protons and neutrons) and leptons (such as electrons). Fermions make up all the matter around us.

- **Bosons:** These particles are like the "glue" that holds everything together. They include photons (which carry the electromagnetic force) and gluons (which hold quarks together inside protons and neutrons). The Higgs boson is another crucial particle, responsible for giving mass to other particles.

Forces of Nature

The Standard Model also explains three of the four fundamental forces that make everything in the universe work:

- **Electromagnetic Force:** This force makes magnets stick to your fridge and keeps electrons orbiting around atoms. It's carried by photons, which act like invisible messages flying between particles.

- **Weak Nuclear Force:** This force plays a key role in certain types of radioactive decay, like the processes that fuel the sun, helping it shine.

- **Strong Nuclear Force:** This is the incredibly strong force that holds the nucleus of an atom together, ensuring that protons and neutrons don't fly apart. It's carried by gluons, which act like superglue.

Why QFT Matters: Real-World Applications

Quantum Field Theory isn't just a collection of abstract ideas—it has real-world applications that help us understand and create things in our everyday lives:

- **Particle Physics:** QFT helps scientists discover new particles, such as the Higgs boson, by predicting where they might be found and how they behave.

- **Materials and Technology:** QFT explains how materials work at the smallest level, leading to discoveries like superconductors (materials that conduct electricity without losing energy) and new electronic devices.

- **Quantum Electrodynamics (QED):** QFT helps us understand how light and matter interact, which is crucial for technologies like lasers and medical imaging.

- **Quantum Chromodynamics (QCD):** This part of QFT explains how the strong nuclear force works, helping us understand the inside of atomic nuclei and leading to advancements in nuclear energy.

Conclusion

Quantum Field Theory might seem challenging at first, but it's incredibly important for understanding how the universe works at its most basic levels. By thinking of particles as ripples in fields and using creative mathematical tools, QFT helps us make sense of everything from the forces of nature to the particles that make up our world. As we continue to explore QFT, we'll keep unlocking new technologies and discovering more about the true nature of reality.

CHAPTER 4

String Theory

Introduction

Imagine shrinking yourself down to the tiniest size possible, smaller than atoms, to see the building blocks of the universe. Surprisingly, you wouldn't find tiny, solid particles. Instead, you'd see something amazing: tiny, vibrating strings, each playing a different note.

According to string theory, these strings are the essence of everything in the universe. Whether it's light traveling across space, the pull of gravity, or the forces holding atoms together, it all comes down to how these strings vibrate and interact.

Think of the universe as a giant orchestra, where each type of particle is like a different note or chord played by these tiny strings. This idea suggests that everything is connected through the vibrations of these fundamental strings.

In this chapter, we'll explore string theory, breaking it down into simple concepts. This "cosmic symphony" might be the key to understanding the mysteries of the universe.

The Basics of String Theory

1. What Are Strings?

In classical physics and quantum mechanics, particles are seen as point-like objects with no size. But string theory suggests that if we could zoom in far enough, we'd see that these particles are actually tiny loops of vibrating strings.

Strings as the Building Blocks: These tiny, one dimensional strings are the most basic elements of the universe. Depending on how they vibrate, they manifest as different particles.

- **Matter:** When strings vibrate in certain ways, they produce particles like quarks and electrons, which make up atoms.

- **Energy:** The vibrations of strings can also represent energy, like photons (particles of light).

- **Force:** The forces that govern interactions between particles, like gravity or electromagnetism, are also described by different vibrational modes of these strings.

In essence, strings are the fundamental "stuff" from which everything—matter, energy, and forces—emerges. The nature of the string's vibration determines what aspect of the universe it becomes.

2. Dimensions

When we think about the world around us, we usually describe it using three familiar dimensions:

1. **Length:** This is how long something is. For example, the length of a pencil from one end to the other.

2. **Width:** This is how wide something is, like the width of a book from side to side.

3. **Height:** This is how tall something is, such as the height of a bottle standing upright.

These three dimensions—length, width, and height—make up the space we move around in every day. They allow us to describe the size and shape of objects and how they relate to each other in space.

What Are Extra Dimensions?

String theory, a cutting-edge idea in physics, suggests that there might be more dimensions beyond the three we're used to. In fact, string theory proposes there could be as many as 10 or 11 dimensions! But if these extra dimensions exist, you might wonder why we don't see or experience them in our daily lives.

The reason we don't notice these extra dimensions is that they're incredibly small and "curled up" in such a way that they're hidden from our view. Think of these extra dimensions as being so tiny that they're beyond the reach of our senses, even with the most powerful microscopes.

An Easy Way to Understand Extra Dimensions

To help understand this concept, let's use a simple analogy involving a piece of paper:

- **From Far Away:** If you look at a piece of paper from a distance, it might seem like just a flat line, especially if you're looking at its edge. In this view, the paper appears to have only one dimension—length.

- **Up Close:** But if you come closer and really examine the paper, you'll notice that it's not just a line. It has a front and back (width) and a thin edge (height). These additional dimensions were always there, but they were hidden from your view when you were far away.

In a similar way, string theory suggests that our world might have more dimensions than we're aware of, but these extra dimensions are so tiny and tightly curled up that we don't notice them in our everyday lives.

Why Are Extra Dimensions Important?

These hidden dimensions aren't just a curious idea—they play a crucial role in string theory. In string theory, everything in the universe is made up of tiny, vibrating strings. The way these strings vibrate determines the types of particles we see and the forces that govern them.

Here's where extra dimensions come into play: the shape and size of these extra dimensions influence how the strings vibrate. Imagine each extra dimension as a tiny, twisted pathway that the strings have to move through. The specific twists and turns of these pathways affect the vibration patterns of the strings, which in turn determine the properties of the particles and forces in our familiar three-dimensional world.

For example, the mass, charge, and other characteristics of particles might depend on how the strings vibrate within these extra dimensions. So, even though we can't see these extra dimensions, they could be shaping the very fabric of our universe in ways we don't yet fully understand.

Understanding extra dimensions helps scientists explore some of the deepest questions about the universe, such as why particles have the properties they do, and how the fundamental forces of nature are connected. While these ideas are still theoretical and haven't been directly observed, they offer a fascinating glimpse into the possible hidden layers of reality that could explain much more about the universe than we currently know.

Supersymmetry: The Hidden Twins of Particles

In the fascinating world of physics, there's a theory called supersymmetry that suggests something intriguing: every particle we know might have a hidden twin that we haven't discovered yet. These twin particles are known as "superpartners," and they could hold the key to solving some of the biggest mysteries in the universe.

Supersymmetric Particles: The Hidden Twins

Let's break it down. Imagine you have a familiar particle, like an electron. According to supersymmetry, this electron has a twin—another particle that shares some similarities but is also different in important ways. This twin is called a "selectron." The selectron would be heavier than the electron, which is one reason why it's been so hard for scientists to find.

These superpartners aren't just copies; they have slightly different properties. For example, while the electron has a negative charge and is light, its twin, the selectron, would be much heavier and might behave differently in certain situations. This difference in weight and behavior is why these superpartners are still hidden from us—they don't interact with the world in the same obvious way as the particles we already know.

Why Haven't We Found These Twins Yet?

Finding these superpartners is like trying to find a needle in a haystack, but with an added twist—the needle is heavier and doesn't react the same way to the magnet you're using to search for it. Because superpartners are heavier and interact less with other particles, they are much harder to detect. Scientists are using powerful machines called particle accelerators, like the Large Hadron Collider, to try and catch a glimpse of these elusive particles. These accelerators smash particles together at incredibly high speeds, hoping to create and observe the superpartners in the process.

Why Supersymmetry Matters

Supersymmetry isn't just a cool idea—it could be a game-changer for our understanding of the universe. Here's why it matters:

- **Unifying Forces:** In physics, there are four fundamental forces—gravity, electromagnetism, and the strong and weak nuclear forces. Supersymmetry could be the missing link that helps scientists combine these forces into a single, unified theory. This would be like discovering that all the different types of weather—rain, snow, wind—are actually part of one big, interconnected system.

- **Dark Matter:** You've probably heard of dark matter—the mysterious stuff that makes up most of the matter in the universe but doesn't emit light, making it invisible to us. One exciting possibility is that some of these hidden superpartners might be the particles that make up dark matter. If this is true, then finding these superpartners could finally explain what dark matter is and how it shapes the universe.

The Big Picture

Supersymmetry is like a grand puzzle that could help scientists unlock new levels of understanding about the universe. While it's still a theory and the superpartners remain hidden, the search continues. If scientists can find these elusive twins, it would not only confirm supersymmetry but also open the door to solving some of the biggest mysteries in physics, from the nature of dark matter to the unification of the fundamental forces.

Understanding the Types of String Theories

Type I String Theory:

Imagine you have a piece of string. You can either leave it straight with two ends, or you can tie the ends together to make a loop. Type I string theory involves both of these possibilities: open strings (with two ends) and closed strings (loops). These strings can vibrate in different ways, and these vibrations are what we think of as particles. This theory also involves both forces and matter, and it's unique because it includes these open strings, which many other string theories don't.

Type IIA and IIB String Theories:

Now let's focus on strings that are always looped, like rubber bands. Type IIA and Type IIB string theories only deal with these looped strings. However, they each have their own special way of looking at the universe. One key difference between them is how they treat extra dimensions—the hidden parts of space that we don't see in our everyday lives. In Type IIA, the extra dimensions are treated in a certain way, while in Type IIB, they are treated slightly differently. These subtle differences help scientists explore different possibilities about how the universe is structured.

Heterotic String Theory:

Imagine you have two different kinds of strings, and each can vibrate in its own way. Heterotic string theory is like combining these different vibrations to explain how particles behave. In this theory, the strings are loops, but they are a bit special. On one side of the loop, the string can vibrate in one way, and on the other side, it vibrates in a completely different way. This combination helps explain some of the complex properties of particles, making it a powerful tool for understanding the universe.

M-Theory:

M-theory is like the grand unifying idea that tries to bring all the different string theories together. It suggests that maybe these different string theories are actually all part of one big picture. But to understand this big picture, we have to think in more than just the usual three dimensions of space and one of time. M-theory uses 11 dimensions—some of which are too small for us to see—to connect the different string theories into one comprehensive theory. This idea is still being explored, but it offers the hope of uniting all the forces and particles in the universe under one umbrella.

Why These Theories Matter:

Each of these string theories offers a different way of understanding the universe. Scientists are like kids with a box of string toys, trying out different ways to see which one best explains the world around us. By exploring these different theories, they hope to find a deeper understanding of how everything in the universe is connected, from the tiniest particles to the largest galaxies.

Why 11 Dimensions?

In string theory, the idea of 11 dimensions is necessary to make all the math and ideas work out neatly.

- **Why Not Fewer Dimensions?** With fewer than 11 dimensions, the theory doesn't explain everything it needs to, like missing pieces of a puzzle.

- **Why Not More Dimensions?** More than 11 dimensions would be like having extra pieces that don't fit anywhere in the puzzle.

- **Why Exactly 11 Dimensions?** 11 dimensions are just the right number to make everything fit perfectly, like having the exact number of puzzle pieces needed to complete the picture.

Why Is String Theory Important?

String theory is more than just an interesting idea—it could help us understand some of the biggest mysteries in science.

1. Unifying All Forces

String theory aims to bring all the forces in nature together into one big picture. Right now, scientists use different theories to explain things like gravity and electromagnetism. String theory suggests that all these forces are just different ways that strings vibrate.

2. Explaining Gravity and Quantum Mechanics

Gravity (which controls big things like planets) and quantum mechanics (which controls tiny things like atoms) don't always agree with each other. String theory might be the way to make them fit together, which would be a huge breakthrough.

3. Understanding Black Holes

Black holes are mysterious objects in space that suck in everything around them, even light. String theory could help us understand what happens inside black holes and how they store information.

String Theory and the Fundamental Forces

String theory promises to unify all four fundamental forces of nature:

1. Gravity and Quantum Mechanics

String theory offers a way to unify gravity, described by Einstein's General Theory of Relativity, with quantum mechanics. In string theory, gravity is mediated by a hypothetical particle called the graviton, which is the result of a specific vibration of a string.

2. The Grand Unification

String theory suggests that all the fundamental forces are different manifestations of the same underlying reality. The vibrations of strings give rise to the different forces, depending on how they vibrate and interact. This is sometimes called the "Theory of Everything."

Conclusion

String theory is one of the most ambitious attempts to understand the fundamental nature of the universe. It offers a framework that could potentially unify all the forces of nature and explain the fabric of reality itself. While it remains a work in progress, with many questions yet to be answered, string theory continues to inspire scientists in their quest for a deeper understanding of the cosmos.

As we move forward, string theory will undoubtedly play a crucial role in shaping our understanding of the universe. Whether it ultimately proves to be the "Theory of Everything" or leads to new insights, it has already expanded our horizons and deepened our appreciation for the intricate and beautiful structure of the universe.

CHAPTER 5

Quantum Consciousness Theories

Introduction

Imagine if the mysteries of the universe and the mysteries of our minds were deeply connected, in ways we've never imagined. What if the strange rules that control the smallest particles in the universe also shape how we think, perceive, and experience reality? This is the fascinating idea behind quantum consciousness theories. These theories suggest that quantum mechanics, the science of the tiniest particles, might hold the key to understanding consciousness—the very essence of our thoughts and awareness. In this chapter, we'll explore this cutting-edge field and see how the quantum world might help us unlock some of the biggest questions about our minds and reality.

What is Consciousness?

Consciousness is the state of being aware of yourself and your surroundings. It's the experience of thinking, feeling, and being aware of those experiences. Even though consciousness is central to our daily lives, it remains one of the greatest mysteries in science. How do physical processes in the brain create the experience of being aware?

The Intersection of Quantum Mechanics and Consciousness

Quantum mechanics is the branch of physics that studies the behavior of the tiniest particles in the universe, like electrons, protons, quarks, and photons. These particles behave in ways that are very different from what we see in our everyday lives. For example, a particle can exist in two places at once, or it can instantly affect another particle far away. Quantum mechanics tries to understand and explain these strange behaviors.

Quantum consciousness theories suggest that these tiny quantum processes might not just happen in space or inside atoms but also inside our brains. These theories propose that the mysterious nature of our thoughts and awareness could be influenced by these quantum processes. In other words, the same rules that govern the smallest parts of the universe might also play a role in how we think, feel, and experience the world.

Key Ideas in Quantum Consciousness

1. Wave-Particle Duality and Consciousness

In quantum mechanics, particles can behave like tiny solid objects or spread out like waves in water. This is called wave-particle duality. Now, some scientists and thinkers have proposed an intriguing idea: What if our consciousness operates in a way that's similar to wave-particle duality? This doesn't mean our thoughts are literally particles or waves, but rather that the principles of quantum mechanics might offer insights into how our minds work.

Just as particles can exist in multiple states at once, our thoughts and decisions might also be in a kind of "superposition"—where we can hold multiple possibilities or options in our minds at the same time. For example, when you're making a decision, you might be considering different outcomes simultaneously, weighing the pros and cons of each choice. In this way, your mind is exploring different "paths" at once, just as a particle explores different states.

The idea here is that consciousness might not be a straightforward, linear process where thoughts follow one after the other in a fixed sequence. Instead, it could be more like a quantum process, where multiple possibilities exist at once, and our awareness "collapses" these possibilities into a single experience or decision—similar to how observing a particle collapses its wave-like behavior into a definite state.

This quantum-like nature of consciousness could help explain some of the complex and sometimes mysterious ways our minds work. For instance, it might offer insights into creativity, where new ideas seem to emerge from nowhere, or intuition, where we suddenly "know" the right answer without consciously going through all the steps to reach it.

2. Superposition and Parallel Processing

Our brains are incredibly complex, and we're often able to think about many different things at once. For example, you might be planning your day, solving a math problem, and remembering a conversation you had earlier—all at the same time.

One way to think about this ability is to compare it to superposition. Just as a quantum particle can exist in multiple states at once, our brains might be capable of holding multiple thoughts or possibilities simultaneously. This doesn't mean that our brains are literally in a superposition like a quantum particle, but it's a helpful analogy to understand how we can process many different ideas at once.

This ability to handle multiple thoughts or tasks at the same time is known as parallel processing. In the context of superposition, you can think of parallel processing as our brain's way of managing different possibilities or solutions simultaneously. Instead of focusing on just one thing and then moving on to the next, our brains can explore different options at the same time.

For example, when you're solving a complex problem, your brain might not just consider one solution at a time. Instead, it might weigh several possibilities together, comparing them and combining different ideas to find the best answer. This process is much faster and more efficient than handling one option after another.

3. Quantum Coherence and Brain Function

Quantum coherence is a remarkable phenomenon in quantum mechanics where particles remain interconnected in a special way, even when they are separated by large distances. This means that the state of one particle can directly influence the state of another, no matter how far apart they are. It's as if these particles are "in sync," working together as one unit, even across great distances.

Imagine two musicians playing a duet on opposite sides of a stage. Even though they're far apart, they're perfectly in tune, playing their notes in harmony. This synchronization is similar to what happens in quantum coherence, where particles behave in a coordinated way, staying connected and influencing each other, regardless of the distance between them.

Some scientists believe that our brains might use a similar type of coherence to create the seamless, unified experience we call consciousness. Our brains are incredibly complex, processing a vast amount of information all at once—what we see, hear, touch, and think. Despite this complexity, our experience feels smooth and continuous, not fragmented or chaotic. How does this happen?

One possibility is that different parts of the brain are able to work together in perfect harmony, much like particles in quantum coherence. This means that, even though different regions of the brain are responsible for different functions (like processing sound, vision, or memory), they might be linked in a way that allows them to communicate and coordinate instantly and seamlessly.

Think about all the things you're experiencing right now—what you see, the sounds around you, the texture of what you're touching, and your thoughts. Each of these experiences is processed by a different part of your brain. Yet, you don't experience them as separate or disjointed; instead, they all come together to form one smooth, unified picture of reality.

This unified experience might be possible because of a form of coherence in the brain. Just as quantum coherence keeps particles linked across distances, coherence in the brain could keep different regions working together in a synchronized way, creating the seamless flow of consciousness.

Key Quantum Consciousness Theories

1. Orchestrated Objective Reduction (Orch-OR)

The Orch-OR theory is an intriguing idea proposed by physicist Sir Roger Penrose and anesthesiologist Stuart Hameroff, which suggests that our consciousness might emerge from quantum processes happening inside tiny structures in our brain cells called microtubules.

Understanding Microtubules

Microtubules are part of the cytoskeleton, the structural framework inside our cells that helps maintain their shape, supports cellular movement, and transports materials within the cell. Imagine microtubules as a network of tiny, hollow tubes that form the scaffolding inside each of our cells, including the neurons in our brain.

Penrose and Hameroff proposed that microtubules might do more than just provide structure and transport. They believe that these microtubules could be capable of performing quantum computations. Quantum computation involves processing information in a way that takes advantage of the principles of quantum mechanics, such as superposition and entanglement, allowing for much more complex and nuanced information processing than classical computation.

In the context of consciousness, the idea is that microtubules might be the site of these quantum computations, and that this quantum processing could be linked to the creation of conscious experience.

Objective Reduction: The Quantum Link to Consciousness

To understand the concept of Objective Reduction (OR), it helps to first know a bit about quantum mechanics. In the quantum world, particles don't have definite properties (like position or state) until they are observed or measured. This means that before observation, a particle exists in a state of possibilities—a superposition—where it can be in multiple states at once.

The Orch-OR theory suggests that within the microtubules, quantum processes are happening, and these processes exist in a superposition of states. Objective Reduction refers to the idea that these quantum processes can't stay in this superposition forever. At some point, they "settle down" or collapse into a single, definite state. Penrose and Hameroff theorize that this collapse of the quantum state within microtubules might correspond to moments of conscious awareness.

In simpler terms, when the quantum computations within the microtubules reach a point where they collapse from a state of possibilities into a definite state, this could generate a spark of consciousness. This process might happen countless times each second, giving rise to the continuous stream of awareness that we experience.

2. Quantum Brain Dynamics (QBD)

Quantum Brain Dynamics (QBD) is a theory that explores how our brain might operate by applying principles from quantum mechanics, suggesting that the brain functions as a unified, interconnected system influenced by quantum processes. This theory aims to explain some of the brain's most complex activities, including consciousness, by looking at how neurons and quantum fields might work together.

Quantum Fields and Neurons: The Brain's Quantum Connection

Neurons are the fundamental cells in our brain that communicate with each other through electrical signals. These signals allow us to think, move, feel, and perform all the functions that define our daily lives. In classical neuroscience, neurons are often thought of as working in a straightforward, linear way—each neuron sending signals to the next, creating chains of communication.

However, the QBD theory proposes a more intricate picture. It suggests that neurons might not just be connected in a simple, one-to-one manner. Instead, they could be linked through quantum fields, which are invisible energy fields that exist throughout the universe. In this context, a quantum field could connect large groups of neurons, allowing them to interact in ways that go beyond the classical understanding of brain function.

If neurons are indeed connected through quantum fields, it means that the brain's activity could be influenced by quantum processes on a much larger scale. This could help explain how the brain manages to perform incredibly complex tasks, like solving problems, making decisions, or experiencing emotions, in a way that feels seamless and unified.

Bose-Einstein Condensates: Collective Behavior in the Brain

One of the more intriguing ideas in QBD involves something called a Bose-Einstein condensate, which is a special state of matter that occurs at extremely low temperatures. In a Bose-Einstein condensate, individual particles lose their separate identities and start to behave as a single, unified entity. This collective behavior allows the particles to act in perfect harmony, almost as if they were one "super-particle."

The QBD theory suggests that something similar might happen in the brain. While our brain is made up of billions of individual neurons, these neurons might sometimes act together in a coordinated, unified way—similar to how particles behave in a Bose-Einstein condensate. This collective behavior could be what allows the brain to handle complex tasks, such as integrating sensory information, forming memories, or generating conscious awareness.

In this view, parts of the brain might operate as a cohesive whole, rather than as a collection of separate neurons. This could be especially important for understanding how consciousness arises, as it suggests that our awareness might not be tied to any one part of the brain, but rather to the coordinated activity of many neurons working together in a quantum state.

3. Holonomic Brain Theory

The Holonomic Brain Theory, proposed by neuroscientist Karl Pribram and physicist David Bohm, presents a fascinating way of thinking about how our brain processes and stores information. This theory suggests that the brain operates much like a hologram, where each part of the system contains the whole picture. This idea challenges the traditional view of the brain as a collection of separate parts each responsible for specific functions.

The Brain as a Hologram: Storing Information Everywhere

To understand the Holonomic Brain Theory, it's helpful to first understand what a hologram is. A hologram is a special type of image that, when illuminated by light, creates a three-dimensional picture. What's unique about holograms is that even if you cut the hologram into smaller pieces, each piece still contains the entire image—just at a lower resolution.

Applying this concept to the brain, the Holonomic Brain Theory suggests that information in the brain isn't stored in one specific location, like a file in a filing cabinet. Instead, information is distributed across the entire brain, much like how a hologram spreads information across its whole surface. This means that each part of the brain holds a representation of the entire set of information.

One of the remarkable implications of this is that if a part of the brain is damaged, the rest of the brain can still access the full information. For example, if a region of the brain associated with memory is injured, other parts of the brain might still retain the ability to retrieve those memories because the information isn't isolated to just one spot.

Implicate Order: The Hidden Reality of the Universe

David Bohm's concept of the "implicate order" is another key element of the Holonomic Brain Theory. The implicate order is a philosophical idea that suggests all the information and structure in the universe is hidden or "folded" into reality in a way that we don't directly perceive. This hidden order contains the potential for everything that happens in the universe, and what we experience in our everyday lives is the "unfolding" of this information into what Bohm called the "explicate order"—the world as we see it.

When applied to the brain, the implicate order suggests that consciousness and thought might arise from this hidden information unfolding within the brain. In other words, our conscious experiences—what we think, feel, and perceive—could be the result of deeper, more fundamental processes that are constantly revealing themselves as the brain interacts with its environment.

Implications of Quantum Consciousness Theories

1. Understanding Consciousness

These theories offer a new way to explain how consciousness—our thoughts, awareness, and sense of self—comes to be, suggesting that quantum processes might play a role in how our minds work.

2. Mind-Body Problem

The mind-body problem is the question of how physical actions in our brain lead to personal experiences, like thinking or feeling emotions. Quantum consciousness theories suggest that our mind might involve deeper quantum processes, which could explain why we experience consciousness.

3. Free Will and Determinism

Quantum mechanics introduces an element of randomness and unpredictability. If these quantum processes happen in our brains, it might mean that we have a form of free will, giving us more control over our actions than we might think.

Connecting the Dots: Understanding Quantum Theories

Here's a simple breakdown of the different quantum theories:

- **Quantum Mechanics:** The basic theory that describes how the smallest particles behave.

- **Quantum Entanglement:** The idea that particles can become linked, so that the state of one affects the other, no matter how far apart they are.

- **Quantum Field Theory:** Treats particles as excitations in underlying fields that fill the universe.

- **String Theory:** Proposes that the building blocks of the universe aren't particles but tiny, vibrating strings.

- **Quantum Consciousness Theories:** Explore whether quantum principles like superposition and entanglement could be linked to consciousness.

43

Conclusion

Together, these concepts suggest that the universe is deeply interconnected, from the tiniest particles to the mysteries of consciousness. They challenge us to rethink our understanding of reality, moving from a world of isolated particles to a web of fields, strings, and possibly even consciousness itself. As we continue to explore these ideas, we get closer to understanding the true nature of the universe and our place within it.

CHAPTER 6

Calculus and Game Theory

Introduction

Calculus and game theory might seem like difficult subjects, but when we break them down, they become powerful tools that help us understand and solve problems in both the natural world and human society. In this chapter, we will explore these two fields in simple terms and see how they connect, with examples that show how they can be used in everyday life.

Part 1: Understanding Calculus

What is Calculus?

Calculus is a branch of mathematics that deals with change. It helps us understand how things move, grow, or shrink over time. The two main ideas in calculus are **differentiation** and **integration**.

Differentiation: Understanding Rates of Change

Differentiation is all about understanding how things change. For example, if you're driving a car, calculus can help you figure out how your speed changes over time.

Example:

Imagine you're driving up a hill and you press the gas pedal a little harder. Your car starts to speed up. Differentiation would help you calculate exactly how fast your speed is increasing. This is useful not just for driving, but also for understanding how populations grow, how prices change, or even how the stock market moves.

Integration: Adding Up Small Parts

Integration is the opposite of differentiation. It's about adding up small pieces to understand the whole picture. If differentiation helps you understand the rate at which something changes, integration helps you understand the total change over a period of time.

Example:

Suppose you're filling a swimming pool with water. Integration would help you figure out how much water you've added to the pool over time. If you know the rate at which water is being pumped in, integration helps you calculate the total volume of water after a certain period.

Why is Calculus Important?

Calculus is important because it helps us understand and predict how things will change. Whether you're an engineer designing a bridge, a biologist studying population growth, or an economist analyzing markets, calculus provides the tools you need to solve complex problems.

Part 2: Introduction to Game Theory

What is Game Theory?

Game theory is a branch of mathematics that studies decision-making. It helps us understand how people, companies, or even countries make choices when they are competing or cooperating with others. The "games" in game theory aren't just about fun—they represent any situation where multiple people or groups are making decisions that affect each other.

The Basics: Players, Strategies, and Payoffs

- **Players:** In game theory, players are the decision-makers. They could be people, companies, or even entire countries.

- **Strategies:** A strategy is a plan of action that a player can take. In any given situation, players choose from a set of possible strategies.

- **Payoffs:** The payoff is the outcome or reward that a player receives based on the strategy they choose and the strategies chosen by others.

Example:

Imagine two businesses competing in the same market. Each business (player) has to decide whether to lower their prices or keep them the same (strategies). The payoff depends on what both businesses decide. If one lowers prices and the other doesn't, the one that lowers prices might gain more customers, but both could end up making less profit if they both lower prices.

The Prisoner's Dilemma: A Classic Example

One of the most famous examples in game theory is the **Prisoner's Dilemma**. Here's how it works:

Two suspects are arrested for a crime. The police don't have enough evidence to convict them on the main charge, so they offer each prisoner a deal:

- If one confesses and the other doesn't, the confessor goes free, and the other gets a heavy sentence.

- If both confess, they both get a moderate sentence.

- If neither confesses, they both get a light sentence.

Each prisoner must decide whether to confess without knowing what the other will do. The dilemma is that while the best outcome for both would be to stay silent, they often both confess, leading to a worse outcome for both. This example shows how individual decision-making can lead to outcomes that aren't the best for everyone involved.

Part 3: Connecting Calculus and Game Theory

Now, you might wonder how these two fields—calculus and game theory—are related. The connection lies in how we can use calculus to analyze and optimize decisions in game theory.

Maximizing Payoffs with Calculus

In game theory, players often want to maximize their payoffs or minimize their losses. Calculus helps in finding the best strategy by showing how payoffs change as players change their strategies.

Example:

Suppose you're a business owner deciding how much to spend on advertising. You want to spend enough to attract customers but not so much that you lose money. Calculus can help you find the "sweet spot" by analyzing how your profit changes with different levels of advertising.

Finding Equilibrium Points

In many games, players reach an equilibrium, where no one can improve their payoff by changing their strategy alone. Calculus can help identify these equilibrium points, showing where players should settle their strategies.

Example:

In a market with two competing businesses, an equilibrium might occur where both set their prices at a level where neither would benefit from changing their prices. Calculus helps determine this price point by analyzing how profits change with different pricing strategies.

Optimizing Long-Term Strategies

Game theory isn't just about one-time decisions. Players often make decisions over time, adjusting as they go. Calculus helps in planning and optimizing these long-term strategies.

Example:

If you're saving for retirement, you might use game theory to decide how much to save each year based on expected returns. Calculus would then help you optimize this strategy, ensuring you reach your savings goals with the least amount of risk or cost.

Part 4: Everyday Applications

Let's look at how calculus and game theory can be applied in everyday life:

- **Personal Finances:** Calculus can help you understand how your savings grow over time with interest, while game theory can guide your decisions on spending and investing, balancing short-term enjoyment with long-term security.

- **Health and Fitness:** Calculus can be used to track your progress in a fitness regimen, understanding how your body changes with different exercise intensities. Game theory might come into play when you decide whether to join a gym, considering both your commitment and potential benefits.

- **Relationships:** Game theory can help you understand social interactions and make better decisions in relationships, whether it's negotiating with a partner or deciding how to resolve a conflict.

- **Career Planning:** Calculus helps in understanding the benefits of further education or training over time, while game theory helps in negotiating salaries or deciding on job offers based on your long-term career goals.

Conclusion

Calculus and game theory might seem like abstract mathematical concepts, but they are incredibly useful tools for solving real-world problems. By understanding rates of change, maximizing payoffs, and finding equilibrium points, we can make better decisions in every aspect of life, from personal finance to career planning. These tools help us navigate a world full of choices and changes, empowering us to take control of our lives and achieve our goals.

Endings as Beginnings: Transitioning from Science to Consciousness

As we reach the end of Module A, we've explored the secrets of the universe, from the building blocks of matter to the mysterious dance of quantum particles. We've seen how science gives us a glimpse into a vast, intricate, and deeply interconnected universe. But these discoveries are just the beginning. They lead us to a fascinating question: Is the universe more than just a collection of atoms and forces? Could the very fabric of reality be intertwined with consciousness? In Module B, we'll go beyond the boundaries of traditional science to explore how these cosmic principles might connect to our own minds, our sense of self, and the possibility that the universe itself is alive with consciousness. Prepare to journey from the physical to the metaphysical, where science meets spirituality, and where the universe might reveal its deepest secrets as a living, breathing entity.

Module B

Thoughts and Reflections

CHAPTER 7

Rethinking the Concept of God

Introduction

Before we dive into this chapter, let's recognize the significance of faith in people's lives. Across the world, billions of people find comfort, guidance, and meaning through their belief in God. For many, this belief is a source of strength that helps them through life's challenges, joys, and uncertainties. God, for many, is the foundation of their lives—a presence that brings hope and stability.

It's not my intention to question or dismiss these deeply held beliefs. I have great respect for the sense of community, comfort, and tradition that faith in God brings to so many. The prayers, rituals, and teachings passed down through generations have shaped human history and culture. They have inspired art, music, literature, and countless acts of kindness. To overlook the significance of God in people's lives would be to ignore a fundamental part of what it means to be human.

However, as we continue this journey—one that seeks to explore the universe and consciousness from a different perspective—I invite you to consider some important questions. These questions might challenge traditional views, but they are essential as we move forward together.

For example, if God is all-powerful and all-knowing, why does suffering exist? Why do natural disasters take innocent lives? Why is there so much pain and injustice in the world? These are not new questions; they have been asked by philosophers, theologians, and everyday people for centuries. Yet, they remain unanswered and continue to trouble many of us.

Moreover, if every religion claims to have the "true" understanding of God, why are there so many different and sometimes conflicting images of God across cultures? How do we reconcile the idea of a loving, caring God with the harsh realities of life? And in a universe as vast and complex as ours, could it be that our traditional ideas about God are limited or even outdated?

These questions aren't meant to create doubt just for the sake of it. Instead, they are an invitation to explore deeper truths, to look beyond the surface, and to seek an understanding of the universe that aligns more closely with what we know from modern science. This chapter isn't about denying God, but about rethinking and redefining our understanding of the divine in a way that resonates with both reason and experience.

As we proceed, I ask for your openness and curiosity. Let's consider the possibility that the universe is more intricate and profound than any single religious narrative can capture. What if, by expanding our view, we could find a sense of the divine that is even more awe-inspiring and interconnected with everything around us?

So, with respect and humility, let's begin this exploration. Together, we'll venture into new ideas, challenge old beliefs, and seek a deeper understanding of the forces that shape our existence. This isn't just a critique—it's an invitation to a broader, richer perspective on life, consciousness, and the universe.

The Problem of Evil and Suffering

One of the biggest challenges to the idea of an all-powerful, all-knowing, and loving God is the existence of evil and suffering in the world.

1. Natural Disasters and Diseases

Natural disasters like earthquakes, hurricanes, and diseases cause enormous suffering and loss of life. These events are often out of our control, making us wonder how a loving God, who cares about humanity, could allow such terrible things to happen. If God is both powerful and loving, why do these devastating events occur?

2. Human-Caused Suffering

Wars, genocides, and acts of violence caused by humans raise difficult questions about divine justice and intervention. If there is a God who has the power to stop these terrible events, why do they still happen? Why does a loving and just God allow people to harm each other in such horrible ways?

3. Innocent Suffering

The suffering of innocent beings, especially children and animals, is a deep moral puzzle. How can an all-loving God allow such pain and injustice to happen to those who have done nothing wrong? The suffering of the most vulnerable challenges the belief in a God who is both all-powerful and completely good.

4. Injustice and Inequality

Religious teachings often portray God as just and fair, but the reality of our world is filled with inequality and injustice.

5. Economic Disparities

There are huge gaps in wealth and resources around the world, leading to poverty, hunger, and a lack of basic necessities for millions of people. If God is truly just, why do these vast inequalities exist? Why do some people live in abundance while others struggle to survive?

6. Social Inequalities

Discrimination based on race, gender, sexual orientation, and other factors is still widespread in many societies. Sometimes, religious teachings have even been used to justify these biases, which makes us question whether these teachings truly come from a just and fair God. How can a God who is supposed to be fair allow such discrimination to persist?

7. Failure to Punish Wrongdoers

Many people who do wrong, such as corrupt leaders and criminals, often avoid punishment and even live successful, comfortable lives. This lack of accountability challenges the belief in a God who is morally righteous and ensures justice. If God is just, why do wrongdoers sometimes seem to thrive while their victims suffer?

Scientific Inconsistencies

Religious beliefs about the natural world often clash with what science tells us.

1. Creation Myths

Many religions have stories about how the universe and life began. However, these stories often don't match up with scientific evidence, like the Big Bang theory or the theory of evolution. For example, while some religious texts say the world was created in a few days, science shows that it took billions of years for the universe to form.

2. Miracles and Supernatural Events

Religions often point to miracles or supernatural events as proof of divine power. But these events usually don't have solid evidence to back them up and can often be explained by natural causes. For instance, what might be seen as a miracle could actually have a scientific explanation that we just don't fully understand yet.

3. Lack of Predictive Power

Scientific theories are powerful because they can make predictions that we can test through experiments. If a scientific idea is wrong, it can be proven false. Religious explanations, however, often rely on faith rather than evidence and can't be tested or proven wrong in the same way, which limits their ability to explain the world around us.

Philosophical and Ethical Issues

Traditional religious ideas about God also bring up deep philosophical and ethical questions.

1. Free Will vs. Predestination

Many religions struggle with the question of whether humans truly have free will if God already knows everything that will happen. If God is all-knowing and has a plan for the future, does that mean our choices are predetermined, and we don't really have free will?

2. Divine Command Theory

Some religious beliefs are based on the idea that what is morally right is determined by what God commands. This can lead to a tricky situation—if something is considered good just because God says so, then morality could seem random or dependent solely on God's will, making it less about right and wrong and more about following orders.

3. The Problem of Divine Hiddenness

If God exists and wants to have a relationship with people, why does God seem so hidden? Why is there so much doubt and so little clear evidence of God's presence? This question, known as the problem of divine hiddenness, challenges the idea of a loving God who wants to be known by everyone.

Alternative Perspectives

Questioning traditional religious ideas about God doesn't have to lead to atheism or the belief that life has no meaning. There are other ways of understanding the universe and our place in it that might make more sense.

1. Pantheism and Panentheism

- **Pantheism:** This belief sees God and the universe as one and the same. Everything that exists—every tree, star, and atom—is a part of God. There's no separation between the Creator and the creation; they are one. This perspective suggests that divinity is present in every part of the universe.

- **Panentheism:** Similar to pantheism, but it adds that God also exists beyond the universe. This means that while everything in the universe is part of God, God is also more than just the physical world.

2. Deism

Deism is the belief that a higher power, often called God, created the universe but doesn't interfere with its workings after the creation. God set the universe in motion with natural laws and then stepped back, letting it operate on its own. Deism contrasts with many traditional religious beliefs that depict God as actively involved in the world and in human affairs.

3. Humanism and Secularism

- **Humanism:** This philosophy focuses on humans and their well-being. It encourages living meaningful lives through compassion, empathy, and mutual respect. Humanists believe that people can solve problems and improve the world using reason, science, and collective effort.

- **Secularism:** This principle promotes the separation of government and religious institutions. Secularism argues that religion should not influence public policy or governance and that people should be free to practice any religion or none at all, without state interference.

Both humanism and secularism emphasize building ethical and moral systems based on human reason, empathy, and the well-being of people, without relying on religious beliefs. These perspectives encourage us to take responsibility for making the world a better place, believing that we have the power to create meaning and justice in our lives.

Conclusion

In this chapter, we've explored some of the challenges that arise when we closely examine the traditional ideas of God presented by various religions. While these beliefs have provided comfort and guidance to many, they also raise challenging questions. The existence of suffering, injustice, and scientific discoveries can make us wonder if our traditional views of God align with the complex and vast world we live in.

However, questioning these traditional ideas doesn't mean giving up on finding deeper meaning or understanding the mysteries of life. Instead, it invites us to broaden our thinking and consider new ways of understanding the divine. By doing so, we can develop a perspective that better fits with what we know about the universe today—a perspective that might offer a more connected and compassionate view of all things.

As we move forward, we'll explore new ideas that attempt to bring together spirituality and science, offering a fresh way to think about our place in the universe. These ideas may challenge what we've always believed, but they also offer the opportunity to deepen our understanding and appreciation of the wonder of existence.

At the end of each chapter, we'll dive into a story that reflects the themes we've discussed. These stories will feature a character facing a significant challenge, one that might seem overwhelming or insurmountable. Through these narratives, we'll explore practical, step-by-step solutions to these problems, offering a clear and actionable path forward. This approach not only helps to illustrate the concepts we've covered but also provides you with real-world strategies that you can apply to your own life.

Story Time: Maya's Journey from Faith to Inner Strength

Maya grew up in a very religious family, where she was taught to believe in a loving and powerful God who was always watching over her. She attended religious services regularly, prayed every day, and believed that her faith would guide her through any challenges in life. But as she grew older, life began to test her beliefs in ways she never expected.

First, Maya lost a close friend in a sudden and tragic accident. The loss felt senseless and cruel, shaking her faith in a kind and protective God. Then, her younger brother was diagnosed with a serious illness, leading to a long and painful struggle. Despite countless prayers, his condition only worsened, leaving Maya feeling abandoned by the God she had always trusted. On top of this, she saw so much injustice in the world—poverty, violence, and corruption—that she began to wonder why a just and powerful God would allow such suffering to continue.

Maya's once strong faith began to crumble. She felt confused, angry, and deeply hurt. The God she had always believed in seemed distant and unresponsive. In her disillusionment, she started to distance herself from her religious practices, feeling lost and disconnected from everything she had once believed.

Problem: Losing Faith in Traditional Beliefs

Maya's disillusionment left her feeling abandoned and disconnected. The suffering she witnessed and experienced, combined with the silence of the God she prayed to, made her question the religious teachings she had followed all her life.

Solution: Discovering a New Perspective

As Maya struggled with her doubts, she began to explore new ideas about spirituality and the universe. She attended a talk on consciousness, where the speaker introduced a different way of thinking about God—not as a being who controls everything, but as a universal consciousness that connects and balances the entire universe.

Questioning Old Beliefs:

Maya realized that the idea of a God who allows suffering didn't make sense to her anymore. She started to question whether traditional concepts of God were created by humans to explain the unknown and to provide comfort. This questioning helped her step back from her pain and consider new possibilities.

Understanding Universal Consciousness:

The concept of a universal consciousness resonated with Maya. She learned that this consciousness is not a person or a deity with emotions, but a fundamental force that keeps the universe in balance. It doesn't intervene in human lives but ensures that the natural laws of the universe are maintained.

Finding Strength in Reality:

Maya began to find strength in the idea that the universe operates according to natural laws, rather than the will of a deity. She realized that suffering is part of life—not a punishment or a test, but a natural consequence of living in a complex world. This understanding brought her a sense of peace and resilience.

Taking Responsibility:

With this new perspective, Maya stopped waiting for divine intervention and started taking responsibility for her own life. She understood that her actions and decisions were powerful and could shape her reality. Instead of relying solely on prayer, she began to actively work on improving her situation and supporting her family.

Redefining Spirituality:

Maya reconnected with spirituality in a way that felt authentic to her. She appreciated the interconnectedness of all things and found comfort in the idea that while there may not be a traditional God, there is a universal consciousness that ties everything together. She also realized that she could still honor the religious traditions she grew up with by seeing them as expressions of this broader, universal consciousness. This new understanding gave her inner strength, resilience, and a sense of purpose grounded in reality.

Over time, Maya's outlook on life transformed. She no longer felt betrayed by a God who didn't answer her prayers. Instead, she embraced a worldview that acknowledged the complexities of life and the power of consciousness, both within herself and in the universe. Maya found peace in accepting life's natural flow and in the power of her own choices to shape her future.

Key Insight:

Maya's story shows how questioning traditional beliefs and exploring new perspectives can lead to personal growth and inner strength. By embracing the idea of a universal consciousness instead of a traditional God, she found a deeper connection to the universe and a renewed sense of purpose. Recognizing that cultural deities can be seen as expressions of this universal consciousness helped her integrate her cultural traditions with a more expansive spiritual understanding.

CHAPTER 8

The Presence of Consciousness in Everything

Introduction

"Everything we call real is made of things that cannot be regarded as real."

– Niels Bohr

As we move away from traditional ideas of God, we find an even bigger mystery: consciousness. What if consciousness isn't just something in our minds but is actually a force that's present everywhere in the universe? This idea suggests that consciousness might not just be in living things but could be a fundamental part of everything around us. This chapter explores the possibility that consciousness, or a form of super-consciousness, connects and balances everything in the universe, both living and non-living.

The Nature of Consciousness

Consciousness is our awareness of ourselves, our surroundings, and our thoughts and feelings. It's what allows us to experience life, make decisions, and understand the world. But even though it's so central to our existence, consciousness is still one of the biggest mysteries in science and philosophy.

- **Philosophical Perspective:**

Philosophers have been debating what consciousness really is for centuries. They ask questions like: Where does consciousness come from? What role does it play in our lives? How does it relate to who we are as individuals? They also explore how consciousness connects to bigger ideas like free will—the ability to make choices—and personal identity—what makes each of us unique.

- **Scientific Perspective:**

Scientists, especially those who study the brain, try to understand consciousness by looking at how the brain works. They study how different parts of the brain interact to create our thoughts, memories, and feelings. By studying the brain's activity, scientists aim to figure out how physical processes in the brain lead to the experiences we have, like seeing a color, feeling happiness, or remembering a past event.

Levels of Consciousness

Consciousness can be thought of as existing on different levels, ranging from basic awareness in simple organisms to the complex thoughts and self-awareness that humans experience.

1. Basic Awareness:

Even the simplest living things, like bacteria or plants, show some form of awareness. They can sense their environment and respond to it. For example, plants grow toward light, and bacteria move away from harmful substances. This kind of awareness is basic, but it's still a form of consciousness.

2. Higher-Order Consciousness:

Humans, and some animals, have a much more advanced form of consciousness. This includes self-awareness, which means being able to think about oneself, and the ability to engage in complex thinking. For example, humans can reflect on their past experiences, plan for the future, and think about their own thoughts and feelings. This higher level of consciousness allows for abstract thinking, like imagining things that don't physically exist or pondering big ideas like love, justice, or the meaning of life.

Universal Consciousness

Some theories suggest that consciousness isn't just something that happens in individual beings but that it might be a fundamental part of the universe itself.

1. Panpsychism

Panpsychism is the idea that consciousness is a universal feature present in all matter, no matter how small. This means that everything in the universe, even the tiniest particles, has some form of consciousness or awareness.

Integrated Information Theory (IIT):

This theory, proposed by neuroscientist Giulio Tononi, suggests that consciousness arises from how well information is integrated within a system. For example, in the human brain, different parts communicate with each other in a highly connected way, which is why we have a high level of consciousness. According to IIT, any system that integrates information in this way, whether it's a brain or something else, could have some level of consciousness.

Implications:

Panpsychism changes the way we traditionally think about mind and matter. Instead of seeing consciousness as something separate from the physical world, panpsychism suggests that consciousness is a natural and essential part of the universe, existing in everything.

2. Quantum Consciousness

Quantum consciousness theories suggest that consciousness is not just a product of brain activity but is deeply connected to the fundamental processes of the universe.

(Note: We already discussed this topic in detail in Chapter 5.)

Orchestrated Objective Reduction (Orch-OR):

This theory, proposed by Sir Roger Penrose and Stuart Hameroff, suggests that consciousness emerges from quantum processes within microtubules—tiny structures inside brain cells. These quantum processes are believed to be linked to the underlying structure of the universe itself, suggesting that consciousness might be connected to a broader, universal consciousness.

Quantum Brain Dynamics (QBD):

This theory views the brain as a system that operates on a quantum level, where quantum coherence helps unify our perceptions and thoughts. This coherence, where particles remain interconnected across distances, could reflect the way consciousness is linked to the super-consciousness of the universe. QBD suggests that our individual consciousness is not separate but is part of this larger, interconnected system, helping us understand how we are tied to the living force that maintains the balance and harmony of the cosmos.

What is Super-Consciousness?

After studying quantum mechanics and string theory, I find myself drawn to the idea that Super-consciousness is a living force that permeates the universe, acting as the medium through which both living and non-living energy interact. This force plays a crucial role in various fundamental theories, connecting and mediating different aspects of reality.

In String Theory: Super-consciousness is the force that connects all strings, ensuring the harmonious interaction of the fundamental building blocks of the universe.

In Quantum Entanglement: Super-consciousness is the underlying information that links entangled particles, enabling them to remain connected regardless of the distance between them.

In Supersymmetry: Super-consciousness bridges the gap between particles and their superpartners, maintaining the symmetry and balance within the universe.

In Quantum Field Theory (QFT): Super-consciousness is proposed as a quantum field itself, a pervasive force that underlies and influences the dynamics of all other fields.

In the Quantum Superposition Principle: Super-consciousness grants quantum systems the ability to exist in multiple states simultaneously until they are observed and collapse into a single state through the act of consciousness.

In essence, super-consciousness is envisioned as the living force that not only sustains but also integrates the diverse elements of the universe, binding everything together into a coherent, interconnected whole. Super-consciousness is the ultimate theory of everything, as it can be explained and understood through multiple foundational theories, serving as the common link that unites them all.

The Role of Super-Consciousness

1. Equilibrium of Strings:

Imagine the universe as a vast symphony orchestra, where each instrument represents a string in string theory. Every instrument (or string) plays its own tune (vibration pattern), and these individual tunes come together to create a harmonious symphony (the universe). The super-consciousness is like the conductor of this orchestra, ensuring that each instrument plays in harmony with the others, maintaining the balance and coherence of the entire performance. If even one instrument falls out of tune, the conductor guides it back, ensuring the symphony continues without disruption. This is how super-consciousness maintains equilibrium among the strings, ensuring that the universe remains stable and coherent.

2. Interconnection:

To understand interconnection, think of how different parts of the orchestra must work together to create beautiful music. The strings, brass, woodwinds, and percussion sections must all be in sync for the performance to sound right. Similarly, in the universe, super-consciousness connects every string to every other string, much like how the conductor ensures that all sections of the orchestra are in harmony. This interconnectedness allows for complex interactions to happen—like the formation of molecules, planets, and even life itself. Without the guiding hand of the super-consciousness, this intricate dance of interactions would fall apart, much like a symphony would collapse into noise without a conductor.

The idea of super-consciousness, as a force that governs the interactions of all particles, reshapes our understanding of reality in a profound way. It suggests that the universe is not just a collection of random, isolated events but a deeply interconnected system, where every part influences and is influenced by the whole. This interconnectedness is not just a physical connection but also a deeper, more profound unity that aligns with ancient spiritual teachings that speak of the oneness of all things.

Example: Think about how ecosystems work. In a forest, the trees, animals, soil, and even the air are all interconnected. The health of the soil affects the trees, which in turn provide shelter for animals, which help spread seeds, and so on. Each element in the ecosystem relies on and supports the others, creating a balanced and thriving environment. Similarly, if super-consciousness governs the universe, it means that every element of reality is similarly interconnected—what happens in one part of the universe can influence what happens in another, no matter how distant or seemingly unrelated.

This idea aligns with philosophical and spiritual traditions like Buddhism and Hinduism, which emphasize the unity of all existence. It suggests that every thought, action, and event is part of a larger cosmic web, all connected through the guiding force of super-consciousness.

Why Is Super-Consciousness a Living Force and Not a Non-Living Entity?

Super-consciousness, when understood as a living force, is different from non-living entities like particles or objects. Unlike non-living forces, which follow fixed laws without awareness or adaptability, super-consciousness operates with purpose, awareness, and the ability to adapt—key characteristics of living forces. Here's how we can understand this concept through real-life examples from different fields:

Healthcare and the Immune System:

Example: When you get a cut, your immune system doesn't just passively respond; it actively detects the injury, sends white blood cells to fight off infection, and begins the healing process. This system operates with a clear purpose and adaptability, showing traits of a living force. Similarly, super-consciousness could be seen as the universal immune system, maintaining the health and balance of the cosmos by guiding and adjusting the interactions of energy and matter.

Environmental Ecosystems:

Example: In a forest ecosystem, different species—plants, animals, insects—interact in a balanced way to maintain environmental health. Predators control prey populations, plants produce oxygen, and animals spread seeds. This system doesn't function by chance; it's a dynamic, self-regulating network that adapts to changes. Super-consciousness can be compared to the force that ensures this balance on a universal scale, guiding the interactions of all things to maintain harmony across the cosmos.

Economics and Market Systems:

Example: In a free market economy, prices and production levels adjust based on supply and demand, maintaining a balance. For instance, if demand for a product increases, the market responds by producing more of it, regulating itself. This self-regulation and adaptability resemble how super-consciousness might guide the universe, ensuring that energy, matter, and events are balanced and in harmony.

Technology and Artificial Intelligence:

Example: Advanced AI systems, like those in autonomous vehicles, make real-time decisions by adapting to their environment. They learn from data, make decisions, and adjust their actions, demonstrating a form of intelligence and purpose. Super-consciousness, in a similar way, could be viewed as the ultimate intelligent system that guides the universe's complex interactions, maintaining balance and order.

Psychology and Human Behavior:

Example: Human consciousness is dynamic and adaptable, constantly evolving based on experiences and environments. We learn, grow, and change, which are traits of a living system. If super-consciousness influences human consciousness, it too would be a living, dynamic force that guides not only individual evolution but the progression of societies and the universe as a whole.

These examples show how systems that are adaptive, purposeful, and self-regulating—key traits of living forces—operate in our world. By comparing these systems to the concept of super-consciousness, it becomes clear that super-consciousness is more than just a non-living entity. It acts as a living force, guiding and maintaining the balance and harmony of all things in the universe, much like how living systems maintain balance and function within their environments. This perspective supports the idea that super-consciousness is an active, living force that plays a crucial role in the ongoing process of the universe.

Conclusion

The idea of consciousness, whether within each of us or as a universal force, presents a powerful alternative to traditional views of God. By considering the possibility of a universal or super-consciousness, we open the door to a deeper understanding of how everything in the universe is connected and governed by unseen forces. This perspective allows us to view the universe not as a collection of random events but as a living, interconnected system. While we still have much to learn and prove, the study of consciousness as a universal element is a fascinating and potentially life-changing field. It holds the promise of transforming our understanding of existence and our place in the universe, offering a more integrated and holistic view of reality.

(Please note that all the above points reflect my personal perspective on Super-consciousness, which I have developed by finding similarities and connections between various proposed concepts and theories in the scientific world. This interpretation is based on how I see these ideas coming together to form a cohesive understanding of the universe's underlying forces.)*

Story Time: Arjun's Journey to Discovering Consciousness

Arjun's Search for Meaning and Connection

Arjun was a successful entrepreneur who had built a thriving business from the ground up. Despite his professional achievements, he often felt empty and disconnected in his personal life. Even though he had accomplished a lot, he couldn't shake the feeling that something was missing. He struggled to form meaningful connections with others and often felt isolated. Arjun was also confused by the different ideas about God he had encountered across various religions, and this only added to his worries about what might happen after death. He believed that his feelings of emptiness were just part of being human and that finding true purpose and connection was something that only a lucky few ever achieved.

Problem: Feeling Disconnected, Confused, and Afraid

Arjun's sense of disconnection, confusion about life's purpose, and fear of death left him feeling constantly empty and dissatisfied. He felt isolated and thought that finding true purpose and meaningful connections was beyond his reach.

Solution: Embracing Universal and Super-Consciousness

Arjun's search for meaning led him to explore new ideas about consciousness. He attended a seminar where the speaker talked about concepts like universal consciousness, panpsychism, and super-consciousness. These ideas struck a chord with Arjun, offering him a fresh way to understand his life and find a deeper sense of connection and purpose.

Discovering Universal Consciousness and Quantum Consciousness

Arjun learned that consciousness might not just be something in our minds but could be a fundamental part of the entire universe. He explored the idea of **panpsychism**, which suggests that consciousness is present in all matter, no matter how small. This idea helped Arjun see his thoughts and experiences as part of a larger, interconnected web of consciousness. He also delved into **quantum consciousness** theories, which propose that quantum mechanics (the science of very small particles) plays a key role in how consciousness emerges. These insights gave Arjun a new perspective on his existence, helping him see his life as part of the universe's intricate design.

Connecting with Super-Consciousness

The concept of **super-consciousness**—the idea that an underlying force maintains balance and harmony throughout the universe—resonated with Arjun. He began to see his life and actions as part of a larger, purposeful pattern. By aligning his thoughts and actions with the principles of super-consciousness, he aimed to create harmony and balance in his life. This shift in perspective helped Arjun find meaning in his daily activities and interactions, seeing them as contributions to the greater order of the universe.

Addressing the Fear of Death

Arjun had always been deeply afraid of what happens after death, which caused a lot of his anxiety. As he explored different spiritual and philosophical ideas, he found comfort in the thought that consciousness might continue in some form after physical death. He also practiced deep meditation techniques that helped him experience a state of consciousness beyond the physical body. Through these practices, Arjun had profound moments of peace and connection, which helped him view death not as an end, but as a transition. These experiences eased his fear of death, as he came to believe that consciousness, in some form, persists beyond individual existence.

Practicing Mindfulness and Meditation

To deepen his connection with universal consciousness, Arjun started practicing mindfulness and meditation every day. These practices helped him become more aware of his thoughts and feelings, allowing him to observe them without judgment and stay connected to the present moment. Through mindfulness, Arjun cultivated a sense of peace and presence, which helped reduce his feelings of isolation and emptiness.

Building Meaningful Connections

Arjun made a conscious effort to build deeper, more meaningful relationships with the people around him. He practiced active listening, empathy, and open communication, recognizing that his relationships were reflections of the interconnectedness of all things. By nurturing genuine connections, Arjun felt more grounded and fulfilled in his interactions.

Reconciling Religious Confusion

As Arjun explored the concept of universal consciousness, he began to reconcile his confusion about the different religious concepts of God. He realized that many religious teachings emphasize the interconnectedness and unity of all things, which aligns with the idea of universal consciousness. Instead of seeing the different religions as conflicting, Arjun started to view them as different interpretations of the same fundamental truth—that consciousness and interconnectedness are at the heart of existence. This realization brought him a sense of peace and acceptance, allowing him to appreciate the spiritual wisdom found in different religious traditions without feeling conflicted.

Over time, Arjun noticed a profound transformation in his life. He felt more connected to others, more aligned with his purpose, and more at peace with himself and the uncertainties of life and death. By understanding and embracing consciousness as a fundamental aspect of the universe, Arjun found the meaning and connection he had been searching for.

Key Insight

Arjun's story shows how exploring the concepts of universal and super-consciousness can help individuals overcome feelings of isolation and find deeper meaning in their lives. By understanding consciousness as an intrinsic part of the universe, we can connect more deeply with the world around us and discover a sense of purpose that goes beyond our individual existence. This understanding can also help ease existential fears and reconcile different religious beliefs, leading to greater inner peace and acceptance of life's mysteries.

CHAPTER 9

Super-Consciousness: The Universal Connection

Introduction

"The universe begins to look more like a great thought than a great machine."

– Sir James Jeans

As we explore the mysteries of the universe, we start to see that everything might be connected by a conscious force—a living awareness that guides and sustains everything. This idea, known as "super-consciousness," suggests that the universe is more than just a collection of objects and events; it might actually be alive and aware in a profound way.

In earlier chapters, we talked about string theory, which says that the basic building blocks of the universe are not tiny particles, but small, vibrating strings. Now, we'll take this idea further by exploring how super-consciousness could be the force that connects all these strings, keeping everything in balance and harmony throughout the universe.

In this chapter, we'll see how super-consciousness fits with string theory and helps us understand the universe as a whole. By thinking of the universe as something alive and conscious, we can start to see the connections between science and spirituality, and how they might work together to explain the world we live in.

The Foundations of String Theory (A Quick Revision)

1. Strings and Vibrations

Imagine that everything in the universe is made up of tiny strings, like super small rubber bands. These strings are so small that we can't see them, even with the most powerful microscopes. Just like a guitar string produces different sounds when plucked in different ways, these tiny strings vibrate in various ways. The way a string vibrates determines what kind of particle it becomes. For example, one type of vibration might create a particle of light, while another might create a particle of matter—the stuff that makes up everything around us.

Types of Strings: There are two kinds of these tiny strings. Open strings have ends, like a piece of string you might find at home. Closed strings are loops with no ends, like a rubber band. These strings don't just exist in the usual three dimensions we know (up-down, left-right, forward-backward). They also exist in additional dimensions that we don't normally see or think about.

2. Extra Dimensions and Calabi-Yau Manifolds

String theory suggests that our universe might have more dimensions than the three we know. Instead of just thinking about things being long, wide, and tall, there could be extra dimensions that are incredibly small and curled up into complex shapes called Calabi-Yau manifolds. These extra dimensions are like tiny hidden spaces where strings vibrate. The shape of these hidden spaces affects how the strings vibrate, which in turn determines the types of particles we get and how the forces in our universe behave.

3. Unification of Forces

One of the main goals of string theory is to explain how all the different forces we know—like gravity and electromagnetism—are really just different vibrations of these tiny strings. For example, string theory suggests that gravity is caused by the vibrations of a special kind of closed string, which is related to a particle called the graviton. This means that all forces in nature might actually be connected, just different vibrations of the same kinds of strings. Understanding this could help scientists explain everything in the universe using a single idea.

Introducing Super-Consciousness

Imagine that the universe is like a giant, complex orchestra, where each instrument (or in this case, each string) needs to play in harmony with the others to create beautiful music. Super-consciousness is like the conductor of this orchestra. The conductor's job is to make sure that every instrument is playing at the right time and in the right way so that the music sounds perfect. Similarly, super-consciousness is an invisible force that ensures all the tiny vibrating strings in the universe work together in harmony. This keeps everything in balance, making sure that the universe stays stable and everything functions smoothly.

1. Equilibrium of Strings

In string theory, all the tiny strings that make up the universe need to vibrate in a way that keeps everything balanced. If one string gets out of sync, it could cause disruptions. Super-consciousness is like the force that makes sure these strings stay in balance, so the universe remains stable and everything in it—like matter and energy—functions as it should.

Harmonizing Vibrations: Just like how a conductor ensures that all the instruments in an orchestra play together harmoniously, super-consciousness ensures that the vibrations of all the strings in the universe are in sync. This creates a balanced and coherent universe, where everything is connected and working together perfectly. Without this harmonizing force, the universe would be chaotic, and nothing would function as smoothly as it does.

Example: The Human Ecosystem

Imagine a small town where everyone has a specific job or role. There are doctors who take care of people's health, teachers who educate children, farmers who grow food, and shopkeepers who sell goods. Each person in the town depends on others to meet their needs, just like the various strings in string theory interact to create the universe.

Now, let's say there's an increase in the town's population. More people mean there's a greater need for food, so the farmers work harder to produce more crops. The shopkeepers stock more goods, and the doctors might need to work longer hours to care for more patients. If any one of these roles becomes unbalanced—for example, if the farmers can't grow enough food—the entire town feels the impact. There might be food shortages, leading to higher prices at the shops, and people might become unhealthy, needing more care from the doctors.

In this scenario, a "super-consciousness" is like the town council that oversees everything. The council doesn't do the work of the farmers or doctors, but it ensures that there are enough resources, that new schools or hospitals are built when needed, and that the town remains in balance. If there's an issue, the council steps in to make adjustments, ensuring that every role is harmonized and that the town functions smoothly.

Similarly, in string theory, super-consciousness ensures that all the "strings" in the universe vibrate in harmony, maintaining balance and stability in the cosmos. Just as the town council keeps everything running smoothly in the town, super-consciousness keeps the universe in equilibrium.

2. Interconnectedness

Super-consciousness acts as an invisible force that ties every string in the universe together, making sure they all interact in a way that creates and maintains the complex systems we see around us—from tiny particles to the entire cosmos.

Universal Web: Imagine the universe as a giant web, with each point where the threads cross representing a string in string theory. These strings are the building blocks of everything, from atoms to galaxies. Super-consciousness is like the glue that keeps this web intact, making sure each string is connected to every other string. This connection allows energy and information to move freely throughout the universe, helping everything work together.

Examples of Interconnectedness:

Connecting Living and Non-Living Things: Think about how plants rely on the sun. The strings that make up a plant are connected to the strings that make up sunlight. Super-consciousness ensures this connection, so when sunlight reaches the plant, it can absorb the energy and grow. This shows how living things (plants) and non-living things (sunlight) are connected, forming a system that supports life.

Connecting Humans with Animals: Humans and animals need the same basic things—food, water, and shelter. The strings in a human body are connected to the strings in an animal's body. For example, when we eat food, whether it's plants or animals, the nutrients become part of our body. This connection shows how energy and matter are shared between humans and other living beings, helping to keep life going.

Connecting Planets and the Sun: The planets, including Earth, orbit the sun in a balanced way, thanks to gravity. In string theory, the strings that make up the sun are connected to the strings that make up the planets. Super-consciousness keeps these strings in sync, which keeps the planets in stable orbits. This connection is important for life on Earth, as it creates predictable cycles like the seasons.

Connecting Life and Death: Life and death are connected through the recycling of matter and energy. The strings that make up a living being are still connected to the strings in the non-living world after death. For example, when a living thing dies, its body breaks down and returns nutrients to the soil, where they help new life grow. Super-consciousness makes sure this cycle continues, keeping life going.

Connecting Humans with the Universe: On a larger scale, the strings that make up stars and galaxies are connected to the strings in our own bodies. Super-consciousness keeps these connections strong, showing that we are part of a bigger system. The elements in our bodies were created in stars long ago, showing that we are deeply connected to the universe.

How Does Super-Consciousness Impart Life to the Universe Through Non-Living Strings in String Theory?

At first glance, it might seem strange to describe super-consciousness as a "living force" when the strings in string theory are considered non-living. However, the concept of super-consciousness goes beyond what we usually think of as "living" or "non-living." Here's how:

1. Super-Consciousness as the Organizer of Strings

In string theory, strings are the fundamental building blocks of the universe, vibrating at different frequencies to create all matter and forces. These strings, on their own, are non-living—they don't have awareness, intent, or life in the way we typically understand it. However, super-consciousness is the force that organizes and harmonizes these strings. It's not that the strings themselves are alive, but that super-consciousness acts upon them, guiding their interactions in a purposeful way, much like how a living being coordinates its actions to maintain balance and achieve goals.

2. Living Force Defined by Function, Not Composition

When we call super-consciousness a "living force," we're talking about how it functions, not what it's made of. This isn't about the physical makeup of strings but about the role that super-consciousness plays—keeping everything balanced, connected, and working together. It's similar to how certain processes in our bodies, like keeping our temperature steady, are crucial for life, even though they're not "alive" themselves.

3. Analogy with Biological Systems

Think about the human body: individual cells or organs aren't "alive" in the sense that they have their own independent consciousness, but the body as a whole functions as a living system. Similarly, super-consciousness can be seen as the guiding principle that brings life-like order to the universe. It operates on a universal scale, making sure that the non-living strings interact in a way that creates the ordered, life-supporting universe we see.

4. Consciousness as a Universal Quality

Finally, if we think of consciousness as a basic part of the universe, then super-consciousness could be seen as the universal expression of this quality. It's not that each string is conscious, but that the whole system of strings, guided by super-consciousness, works together in a way that reflects a living, dynamic process—just like how neurons in the brain create thought.

So, when we talk about super-consciousness as a living force, we don't mean that the strings in string theory are alive. Instead, we mean that this force gives life-like order, balance, and purpose to the universe. It acts as the guiding energy that connects and harmonizes everything, turning the non-living strings into a dynamic, life-supporting system. In this way, super-consciousness can be understood as a living force that orchestrates the dance of the cosmos.

Implications of Super-Consciousness in String Theory

1. Unified Framework

The concept of super-consciousness provides a unifying idea that could bring together string theory and our understanding of the universe in a more complete way.

Holistic Perspective: When we think of the universe as an expression of super-consciousness, it helps us see that everything is connected and works together. Just as the strings in string theory are connected and influence each other, super-consciousness helps us understand that all parts of the universe, from the smallest particles to the largest galaxies, are part of a single, interconnected whole. This perspective emphasizes that nothing exists on its own; instead, everything in the universe is part of a larger, connected system.

2. Quantum Mechanics and Consciousness

Super-consciousness could help explain the mysterious connection between quantum mechanics (the science of the very small) and consciousness (our awareness and thought processes).

Quantum Coherence: In quantum mechanics, particles can exist in multiple states at once and can be connected across distances—a phenomenon known as quantum coherence. Super-consciousness might be the force that keeps this coherence, especially in living systems like the human brain. This could mean that super-consciousness helps organize and maintain the complex quantum states needed for consciousness to emerge. Essentially, super-consciousness could be the glue that holds together the delicate balance of quantum processes in our brains, leading to our experience of awareness and thought.

3. Fundamental Forces and Consciousness

Super-consciousness might also provide new insights into how fundamental forces, like gravity, are connected to consciousness.

Graviton and Consciousness: In string theory, gravity is thought to be controlled by a particle called the graviton. Super-consciousness could influence how gravitons interact with the fabric of spacetime, possibly linking gravity to conscious processes. This suggests that consciousness might not just be a product of biological processes in the brain, but could also be influenced by the fundamental forces that shape the universe. If super-consciousness affects how gravity works, it could mean that our thoughts and experiences are connected to the very structure of reality itself.

How String Theory and Super-Consciousness Answer Questions About Reality, Free Will, and the Mind

String theory, with its idea of tiny vibrating strings as the building blocks of the universe, offers an interesting way to think about some big questions. When we add the idea of super-consciousness into string theory, we get a deeper understanding of reality, free will, and the mind. Here's how:

1. **Understanding Reality:** In string theory, all particles and forces come from the vibrations of tiny strings. When we think of super-consciousness as the force that harmonizes these vibrations, it suggests that reality is not just a collection of random events. Instead, reality is like a symphony, where every string (or particle) plays a part in creating a coherent whole. This view implies that reality is more than just physical objects and forces—it's also shaped by a deeper, conscious force that gives purpose and order to everything in the universe.

2. **Free Will:** String theory tells us that the universe is interconnected at the most basic level, with every string influencing every other. When we add super-consciousness into the mix, it suggests that our ability to make choices—our free will—is also part of this interconnected system. While we have the freedom to make decisions, these decisions are woven into the broader fabric of the universe. This means that our choices are not isolated; they are part of a larger, conscious plan that links us to the entire cosmos.

3. **The Mind:** String theory's idea of interconnected strings, when combined with super-consciousness, helps us see the mind as more than just the activity of neurons in the brain. It suggests that our thoughts, emotions, and experiences are connected to a universal consciousness that permeates the entire universe. This connection means that our minds are not just individual, isolated entities—they are part of something much bigger, linking us to the cosmos itself. This perspective helps explain how our minds can influence reality and how the universe, in turn, influences our thoughts and experiences.

In simple terms, by combining string theory with the concept of super-consciousness, we can think of the universe as a living, interconnected system. In this system, reality, free will, and the mind are all intertwined, working together as part of a larger, conscious design.

Religious God vs. Super-Conscious Universe: Two perspectives of the Same Story

Religious God Perspective:

The Wright brothers, Orville and Wilbur Wright, made their first successful powered flight on December 17, 1903. This historic event took place near Kitty Hawk, North Carolina, and is considered the first time a controlled, sustained flight of a powered, heavier-than-air aircraft was achieved. Their airplane, called the **Wright Flyer**, flew for 12 seconds and covered a distance of 120 feet during its first flight. The Wright brothers' first flight may have lasted only 12 seconds, but it set in motion a series of events and innovations that have profoundly shaped the modern world.

From a traditional religious viewpoint, one might say that God guided the Wright Brothers in their invention of the airplane. This idea suggests that God chose to help Orville and Wilbur Wright achieve their dream of flight, providing them with the knowledge and perseverance they needed. However, this perspective brings up some difficult questions:

1. **Why Only Them?** If God was helping the Wright Brothers, why did He choose them over other inventors or people who might have needed divine help more, like those suffering from illness or poverty? This raises the question of whether God plays favorites, and if so, why.

2. **Good and Bad Uses:** The airplane has been used for both positive purposes, like connecting people across the world, and negative ones, like warfare. If God inspired the invention of the airplane, why would He allow it to be used for harm as well as good? This makes it hard to understand the moral reasoning behind divine intervention.

3. **Inconsistent Intervention:** If God intervenes in human affairs to help invent something like the airplane, why doesn't He also intervene to stop disasters or prevent suffering? The selective nature of this intervention can be confusing and raises questions about fairness.

This perspective can make it challenging to understand how and why God might choose to intervene in the world, especially when considering the mixed outcomes of human inventions.

Super-Conscious Universe Perspective:

Now, let's look at the Wright Brothers' achievement from the viewpoint of a super-conscious universe. Here's how this idea makes sense of their invention:

1. **Their Decision:** The Wright Brothers made a conscious choice to pursue the dream of flight. This was their personal goal, driven by curiosity and determination.

2. **A Bigger Plan:** Their success didn't just happen in isolation. It came at a time when the world was becoming more connected through technology and industry. The invention of the airplane met a growing need for faster travel and communication, fitting into a broader pattern of human progress.

3. **Super-Consciousness in Action:** According to the idea of super-consciousness, the Wright Brothers' decision was part of a larger, universal plan. Their invention wasn't just about achieving personal success; it was about fulfilling a deeper need within the universe for innovation and connection. Their work aligned with the broader goals of humanity, making their success part of a bigger picture.

4. **Impact Beyond Their Lives:** The airplane changed the world in ways the Wright Brothers probably never imagined. It revolutionized travel, trade, and even influenced global events like wars. This ripple effect shows that their invention wasn't just about them—it played a crucial role in the ongoing story of human development.

5. **Universal Connection:** The Wright Brothers' story shows how individual choices can connect with and contribute to the larger flow of the universe. Their invention wasn't just a personal achievement; it was part of a bigger, interconnected system where each action influences the whole.

Super-consciousness allows the airplane to be used for both positive and negative purposes because it operates within a framework of free will and human responsibility. It doesn't dictate how inventions are used but provides the space for human choice. The dual nature of the airplane—as a tool for both connection and conflict—reflects the broader principle that humans must learn to navigate the ethical implications of their creations. This process is essential for growth and evolution, as it challenges humanity to strive for balance and make choices that ultimately align with the greater good.

In this perspective, the Wright Brothers' success reflects a connection with a broader, universal consciousness. Their decision to invent the airplane wasn't just a personal choice—it was part of a larger plan that shaped the course of history, showing how individual actions can align with the universe's deeper goals and lead to significant progress for all of humanity.

Conclusion

In this chapter, we've explored the concept of super-consciousness and its deep connection to string theory. By viewing the universe as a living, interconnected system, where every string is harmonized by a conscious force, we gain a deeper understanding of reality, free will, and the mind. Super-consciousness offers a unifying framework that ties together the seemingly different elements of the cosmos, revealing the intricate web of connections that bind everything from the smallest particles to the largest galaxies.

Through this lens, the universe is not a random collection of events, but a coherent and purposeful whole, where every action, thought, and event is part of a larger, conscious design. This perspective not only bridges the gap between science and spirituality but also provides a more complete understanding of our place in the universe.

As we continue to explore these ideas, we open ourselves to a richer, more integrated view of existence—one that recognizes the interconnectedness of all things and the presence of a guiding, conscious force that shapes the fabric of reality. This journey into the nature of super-consciousness invites us to rethink our understanding of life, the universe, and the role we play within this grand, cosmic symphony.

Story Time: Asha's Relationship Struggles

Asha was a talented software engineer who had worked hard to build a successful career. However, her personal life was not as successful. She found herself facing repeated failures in her romantic relationships. No matter how much effort she put into them, her relationships always seemed to fall apart. They often started well but ended due to misunderstandings, emotional distance, and breakups. Asha began to believe that maybe she was just bad at relationships or that she was meant to be alone. She thought others were better at relationships because they had better communication skills or just got lucky in finding the right partners.

The Problem:

Asha's struggle was that she couldn't form and keep deep emotional connections in her relationships. This led to repeated failures, and she started to believe that successful relationships were all about luck or other people being better at it.

The Solution:

One day, Asha decided she needed to understand why her relationships were failing. She attended a mindfulness retreat where she learned about the concepts of string theory and super-consciousness. These ideas suggest that everything in the universe is connected, much like strings that vibrate and interact with each other.

Understanding Connection in Relationships:

Asha began to see her relationships as part of a larger, interconnected web of experiences and emotions. She realized that her thoughts, emotions, and actions were all connected, and that her previous approach of keeping her emotions separate and focusing only on the practical side of a relationship was preventing her from truly connecting with her partners.

Identifying Emotional Blocks:

Through self-reflection and mindfulness practices, Asha discovered that her fear of being vulnerable and her belief that showing emotions was a sign of weakness were the main reasons she struggled to connect with others. She realized that hiding her feelings and keeping an emotional distance was causing her relationships to fail.

Changing Beliefs and Embracing Vulnerability:

Asha decided to change her beliefs about emotions and vulnerability. She started to see emotional connection as a strength, not a weakness. By allowing herself to be open and vulnerable, she began to create deeper and more meaningful connections with her partners.

Becoming More Emotionally Aware:

Asha started practicing mindfulness every day to become more aware of her emotions and how they affected her interactions with others. She made an effort to be fully present in her conversations, listen to her partner's feelings, and express her own emotions honestly. This helped her build stronger emotional connections.

Improving Communication and Building Trust:

Asha also worked on her communication skills. She learned to clearly express her needs and boundaries while also being open to her partner's needs. This honest and open communication helped her build trust and create a strong foundation for her relationships.

Aligning with Super-Consciousness:

Asha began to see her relationships as part of a larger, interconnected whole, guided by the principles of super-consciousness. She understood that by focusing on balance, harmony, and mutual respect, she could create healthier and more fulfilling relationships.

The Outcome:

Over time, Asha noticed significant improvements in her relationships. She was able to form deeper emotional connections, communicate more effectively, and maintain a healthier balance between her personal and professional life. Asha realized that by understanding the interconnectedness of relationships and developing emotional awareness, she could overcome her previous struggles and build lasting, meaningful connections.

Key Insight:

Asha's story teaches us that our relationships are shaped by everything we think, feel, and do. By understanding how everything is connected, becoming more aware of our emotions, and embracing vulnerability, we can overcome relationship challenges and create deeper, more meaningful connections. Asha's journey shows how applying these principles can lead to personal growth and success in relationships.

CHAPTER 10

Karma and the Quantum Flow

Introduction

"In the universe, everything is connected; there is no such thing as chance."

– Carl Jung

Karma, a concept deeply rooted in Eastern philosophies, teaches us that every action we take has consequences, shaping our future in ways we might not immediately see. Traditionally, karma is viewed as a spiritual or moral law, guiding our lives through a cycle of cause and effect. But what if we looked at karma through the lens of super-consciousness—the idea that everything in the universe is connected by a living, conscious force? In this chapter, we'll explore how karma and super-consciousness work together, showing how our actions ripple through the universe, influencing the delicate balance of energy and matter. By understanding this connection, we can gain a deeper insight into how our choices shape not only our own lives but the very fabric of the cosmos.

Understanding Karma

1. Traditional Perspectives

In many Eastern philosophies, karma is viewed as a moral principle where good actions lead to positive outcomes and bad actions lead to negative consequences. This concept is often closely tied to the idea of rebirth and spiritual evolution.

- **Hinduism and Buddhism:** In Hinduism, karma is a core concept that influences one's future lives and spiritual progress. For example, in the Hindu epic "Mahabharata," the character Yudhishthira is often depicted as someone who adheres strictly to dharma (righteous duty), understanding that his actions in this life will affect his soul's journey in future incarnations. Similarly, in Buddhism, karma is seen as a law of cause and effect that influences one's path to enlightenment. The "Dhammapada," a collection of sayings attributed to the Buddha, emphasizes that our actions, words, and thoughts shape our future experiences. For instance, the verse "All that we are is the result of what we have thought" reflects the Buddhist understanding of karma as an ever-present force that guides the course of our lives based on our deeds.

2. Modern Interpretations

Modern interpretations of karma have expanded beyond just moral actions to include the broader consequences of all actions, including thoughts and intentions.

- **Universal Law:** Karma can be understood as a universal law that ensures every action, whether physical, verbal, or mental, has an equal and opposite reaction. This idea is reflected in various works of literature and philosophy. For example, in "The Bhagavad Gita," Krishna advises Arjuna that actions performed with selfless intentions and detachment from outcomes contribute to positive karma. This interpretation of karma suggests that it's not only the action itself but the motivation behind it that determines the karmic result. In modern times, this perspective is echoed in self-help books like "The Secret" by Rhonda Byrne, which suggests that our thoughts and intentions can manifest in our lives through the law of attraction, a concept closely related to karma.

By examining karma through both traditional and modern lenses, we see that it is not just a spiritual or moral concept but a universal principle that operates in all aspects of life, ensuring that every action we take, whether good or bad, has a corresponding consequence.

Super-Consciousness and Karma

The concept of super-consciousness offers a scientific and philosophical framework for understanding how karma functions at the most fundamental levels, particularly through the lens of string theory and quantum physics.

It connects every action we take to the intricate balance of energy and matter in the universe.

1. Interconnectedness of Actions

In string theory, every particle and force in the universe arises from the vibrations of tiny strings, which are all connected within a multi-dimensional space. Super-consciousness, as the guiding force behind these strings, ensures that our actions are not isolated incidents but are instead woven into the fabric of the universe.

- **Energy Transfer in Quantum Terms:** In quantum physics, particles can exchange energy through interactions, and these exchanges can have far-reaching effects. Similarly, our actions can be viewed as creating ripples of energy within the universe's web of vibrating strings. Positive actions might align these strings in a way that supports constructive energy patterns, contributing to harmony and coherence within the system. On the other hand, negative actions could introduce disturbances, creating chaotic or destructive energy patterns that disrupt this harmony.

2. Equilibrium and Balance

Super-consciousness functions as the force that maintains the equilibrium of the universe by balancing the energy and matter produced by our actions. This balancing act can be seen as a quantum-level process that reflects the principle of karma.

- **Restoring Balance in the Cosmic Web:** When an action disrupts the balance of energy—akin to introducing a disturbance in the delicate vibrational patterns of strings—super-consciousness acts to restore this balance. In quantum physics, when a system is disturbed, it naturally seeks a new state of equilibrium. Similarly, super-consciousness adjusts the interactions and vibrations of strings to correct any imbalances caused by actions, thereby ensuring the universe remains stable and coherent. This process of restoring balance is the mechanism behind karmic consequences, where the universe responds to actions in a way that reestablishes harmony and order.

In this view, karma is not merely a moral or spiritual concept but a reflection of the natural laws that govern the universe at the quantum level. Super-consciousness, through its influence on the strings that form the foundation of reality, ensures that every action has a corresponding reaction, maintaining the balance and integrity of the cosmos.

Historical Example: The Rise and Fall of Civilizations as a Reflection of Karma and Super-Consciousness

The rise and fall of ancient civilizations offer a compelling real-world example of how the concept of karma and super-consciousness can be understood through the lens of string theory and quantum physics.

1. The Roman Empire: Positive Actions and Constructive Energy Patterns

At its height, the Roman Empire was a powerful force, known for its advancements in law, engineering, and governance. The positive actions of the Roman civilization, such as the establishment of laws (like the Twelve Tables) and the creation of infrastructure (such as roads and aqueducts), can be seen as creating constructive energy patterns. These actions aligned with the principles of super-consciousness, contributing to the stability and growth of the empire. The interconnectedness of their actions, such as building trade routes and fostering cultural exchange, allowed the empire to thrive, with energy flowing harmoniously within the web of their society.

- **Energy Transfer and Constructive Patterns:** The Roman Empire's constructive actions, like expanding trade networks and developing advanced engineering techniques, created positive ripples within the interconnected web of their civilization. These actions resonated through time, contributing to the empire's strength and cohesion, similar to how positive vibrations in string theory maintain harmony within a system.

2. The Fall of Rome: Negative Actions and the Disruption of Balance

However, over time, the Roman Empire also engaged in negative actions—political corruption, overexpansion, and the exploitation of resources and people. These actions disrupted the balance within the empire and led to the eventual collapse. From the perspective of super-consciousness, these negative actions introduced destructive energy patterns, creating disharmony within the empire's societal structure.

- **Restoring Balance:** As the empire's actions became increasingly unbalanced, super-consciousness, working through the interconnected web of their society, sought to restore equilibrium. This is reflected in the decline of Rome, where internal strife, economic instability, and external pressures gradually led to the empire's downfall. The fall of Rome can be seen as the universe's way of restoring balance after a period of significant disruption, aligning with the concept of karmic consequences.

3. The Aftermath: Rebalancing through the Middle Ages

After the fall of the Roman Empire, Europe entered the Middle Ages, a period often seen as a time of rebalancing. The decentralized political structures, the rise of feudalism, and the spread of new religious and philosophical ideas can be viewed as the universe seeking a new state of equilibrium. The decline of centralized power and the emergence of new social orders represented a shift towards a more balanced, if fragmented, system.

- **Super-Consciousness in Action:** The transition from the Roman Empire to the Middle Ages illustrates how super-consciousness operates to maintain the balance of energy in the universe. The collapse of a powerful but increasingly corrupt and unstable empire was followed by the emergence of new structures that allowed for a different, perhaps more sustainable, form of societal organization.

This historical example demonstrates how the rise and fall of civilizations can be understood as reflections of karmic principles, with super-consciousness working to maintain balance and harmony within the universe. Just as in string theory, where disruptions are corrected to restore equilibrium, the universe responds to the actions of civilizations—rewarding constructive energy patterns and correcting destructive ones.

Mechanisms of Karma in the Universe

1. Quantum Mechanics and Karma

Quantum mechanics, which deals with the behavior of particles on the smallest scales, offers a unique perspective on how karma might work in the universe.

- **Entanglement:** In quantum physics, entanglement is a phenomenon where particles become linked, so that the state of one particle instantly affects the state of another, no matter how far apart they are. When we apply this concept to karma, it suggests that our actions create connections—entangled states—that influence not just our immediate surroundings, but potentially far-reaching parts of the universe. These entangled connections mean that the consequences of our actions ripple out, affecting other people, situations, and even the broader fabric of reality.

- **Superposition:** Superposition in quantum mechanics means that particles can exist in multiple states at once until they are observed or measured. This can be likened to the potential outcomes of our actions existing in a superposition of possibilities. Until a decision or action is made (which collapses the superposition into a single outcome), the future remains open with multiple possibilities. This idea connects to karma by suggesting that our actions shape these possibilities, influencing what eventually happens in our lives and the world around us.

2. String Theory and Karma

String theory, which proposes that everything in the universe is made up of tiny, vibrating strings, provides another layer of understanding for karma.

- **Vibrational Patterns:** In string theory, different particles and forces are the result of different vibrational patterns of these fundamental strings. When we think of karma in this context, our actions could be seen as affecting the vibrational state of these strings. Positive actions might create harmonious vibrations, leading to balance and well-being in the universe, while negative actions could disrupt these vibrations, causing disharmony and negative consequences. This perspective suggests that every action we take has the potential to influence the underlying vibrations that make up reality, affecting not just ourselves but the entire cosmos.

Karma and Consciousness

The connection between karma and consciousness is vital for understanding how our thoughts, intentions, and actions contribute to the karmic outcomes we experience.

1. Intentions and Energetic Imprints

Conscious intentions can be thought of as creating energetic imprints that interact with the universe on a fundamental level.

- **Thought Forms:** In the framework of quantum physics and string theory, our thoughts and intentions might generate energy that interacts with the web of vibrating strings that make up reality. These thought-generated energies can create positive or negative imprints within this web. Positive intentions might align the strings in a harmonious way, leading to beneficial outcomes, while negative intentions could create distortions or disruptions, leading to challenges or difficulties as karmic consequences.

2. Mind-Body Connection

The connection between our mind and body illustrates how our internal states of consciousness influence our external actions and their karmic results.

- **Holistic Influence:** Just as quantum mechanics suggests that observing a particle can influence its state, our conscious awareness and mindfulness can guide our actions in a way that creates positive karmic energy. By maintaining a balanced and mindful state of consciousness, we can ensure that our actions align with the harmonious vibrations of the universe, fostering balance and harmony not only within ourselves but in the broader web of reality. This holistic approach highlights that by being aware of our thoughts and actions, we can influence the karmic outcomes we experience, leading to a more balanced and fulfilling life.

Practical Implications of Karma and Super-Consciousness

1. Ethical Living

Understanding karma through the lens of super-consciousness and string theory encourages us to live ethically, recognizing that our actions have far-reaching effects on the universe.

- **Mindful Actions:** Just as the vibrations of strings in string theory can influence the fabric of reality, our actions can create ripples that affect not just our own lives but the world around us. By being mindful of our actions and their consequences, we can contribute to a positive balance of energy in the universe. This mindfulness can lead to better physical and mental health, as positive actions and thoughts align us with the harmonious vibrations of the universe, reducing stress and promoting well-being.

2. Personal Responsibility

The idea of karma and super-consciousness emphasizes the importance of taking responsibility for our actions, understanding that everything we do impacts the larger web of existence.

- **Accountability:** In string theory, every string is connected, meaning that what happens to one string can affect others. Similarly, our actions are interconnected with the broader universe. Recognizing this interconnectedness fosters a sense of accountability, encouraging us to act with integrity and compassion. This perspective can help solve relationship issues by promoting understanding and empathy, leading to healthier and more fulfilling connections with others.

3. Spiritual Growth

Viewing karma through the lens of super-consciousness and quantum mechanics supports spiritual growth and self-awareness, guiding us on a journey of personal and collective evolution.

- **Inner Harmony:** In quantum physics, the state of a particle can influence the state of another, even across great distances, a concept known as entanglement. Similarly, our actions and intentions can influence the energy around us. By aligning our actions with positive intentions, we promote inner harmony, which in turn contributes to the overall balance of the universe. This alignment can lead to greater success in all areas of life, including wealth, as a balanced and harmonious inner state attracts positive opportunities and outcomes.

Solving Life's Problems with Karma and Super-Consciousness

By applying the principles of karma and super-consciousness, we can address and solve many of the challenges we face in life:

- **Physical and Mental Health:** Ethical living and mindful actions can reduce stress and promote overall well-being, leading to better physical and mental health. When we align with the positive vibrations of the universe, our bodies and minds naturally move towards a state of balance and health.

- **Relationships:** Understanding the interconnectedness of all things helps us build stronger, more empathetic relationships. By taking responsibility for our actions and being mindful of their impact, we can create deeper, more meaningful connections with others.

- **Wealth:** When we align our intentions with positive energy and act with integrity, we open ourselves to opportunities for growth and success. The universe, seen through the lens of super-consciousness, rewards those who contribute positively to its balance, leading to increased wealth and prosperity.

In conclusion, by living in accordance with the principles of karma and super-consciousness, we can create a life that is healthier, more connected, and more prosperous. These concepts remind us that our thoughts, actions, and intentions are all part of a larger, interconnected system that influences every aspect of our lives.

Conclusion

When we look at karma through the lens of super-consciousness, we gain a deeper understanding of how our actions influence the energy and matter in the universe. String theory and quantum physics show us that everything is interconnected, and super-consciousness works to maintain balance in this web of connections. By being aware of this, we realize the importance of living ethically, taking responsibility for our actions, and growing spiritually.

This perspective encourages us to align our actions with the natural order of the universe, helping us create a life filled with peace, health, strong relationships, and true happiness. While scientific proof of these ideas is still evolving, exploring karma and super-consciousness offers us valuable insights into how the universe works and how we can live in harmony with it.

Story Time: Rohit's Journey from Rebellion to Growth

Rohit was a teenager who often felt misunderstood by his parents and teachers. Frustrated and resentful, he began to rebel against authority—skipping classes, ignoring homework, and getting into trouble with his peers. As his grades slipped, he became more isolated from his family and friends. Rohit felt that life was unfair and believed he was just unlucky. He convinced himself that others were more successful or well-behaved because they had better circumstances, more supportive parents, or simply better luck. "Why should I even try when everything is stacked against me?" he thought.

Problem: Struggles During Adolescence Due to Rebellion

Rohit's rebellious behavior led to poor academic performance, strained relationships with his family and teachers, and a growing sense of frustration and hopelessness. His belief that his circumstances were due to bad luck or karma prevented him from recognizing how his own actions were contributing to his problems.

Solution: Understanding Karma and Taking Responsibility for Actions

Rohit's downward spiral continued until one day, after another heated argument with his parents, his grandfather decided to have a serious talk with him. His grandfather explained the concept of karma—not as a mystical force, but as a natural consequence of one's actions. He told Rohit that karma is about how our choices and behaviors create ripple effects that shape our future. He emphasized that Rohit had the power to change his path by changing his actions and taking responsibility for his life.

1. **Becoming Aware of Consequences and Karma**

   Rohit began to reflect on his grandfather's words. He realized that his poor grades and strained relationships weren't just the result of bad luck—they were the direct consequences of his choices. Skipping classes and ignoring responsibilities were creating negative outcomes, or "bad karma," leading to the very problems he was frustrated with. This realization helped Rohit understand that he was not a victim of circumstance, but an active participant in creating his reality.

2. Identifying Emotional Triggers and Root Causes

Through introspection, Rohit identified the emotional triggers behind his rebellious behavior. He recognized that much of his rebellion stemmed from feelings of inadequacy and a desire to prove himself. These feelings had led him to reject authority and make choices that ultimately harmed him. By addressing these root causes, Rohit began to see that his rebellion was not helping him and was instead reinforcing the negative outcomes he wanted to escape.

3. Taking Responsibility and Making Conscious Choices

Motivated by this new understanding, Rohit decided to take responsibility for his actions. He started attending classes regularly and made an effort to complete his assignments on time. He also began repairing his relationships with his family and teachers by communicating more openly and showing respect. Rohit understood that by making positive, conscious choices, he could create better outcomes for himself.

4. Reframing Thoughts and Embracing a Growth Mindset

Rohit shifted his mindset from one of helplessness to one of empowerment. Instead of thinking, "I'm unlucky, and things will never change," he began to tell himself, "I have the power to change my life through my actions." He embraced a growth mindset, believing that he could improve and succeed if he put in the effort. This new perspective helped Rohit stay motivated and focused on his goals.

5. Practicing Mindfulness and Staying Present

To stay on track, Rohit incorporated mindfulness practices into his daily routine. By staying present and aware of his thoughts and actions, he was able to catch himself before falling back into old patterns of rebellion. Mindfulness helped him stay focused on the positive changes he was making and reinforced his commitment to creating better outcomes.

6. Building Positive Relationships and Seeking Support

Rohit also realized the importance of surrounding himself with supportive people who encouraged his growth. He reconnected with friends who were focused on their studies and started building positive relationships with his teachers. By seeking support from others and maintaining these positive connections, Rohit found it easier to stay on the right path.

Over time, Rohit noticed significant improvements in his life. His grades began to rise, his relationships with his family improved, and he felt a renewed sense of purpose and direction. Rohit realized that by understanding karma and taking responsibility for his actions, he could change his reality and create a better future for himself.

Key Insight:

Karma is not just a mystical force but a logical outcome of our actions. By understanding that our choices create ripple effects that shape our future, we can take responsibility for our lives and make conscious decisions that lead to positive outcomes. Rohit's story illustrates how embracing the concept of karma and making intentional, responsible choices can help overcome the challenges of adolescence and lead to personal growth and success.

CHAPTER 11

Thoughts as Quantum States

Introduction

"The more I study science, the more I believe in God."

– Albert Einstein

The human mind is one of the most complex and mysterious aspects of our existence. The process by which our physical brain generates thoughts and consciousness has puzzled scientists and philosophers for centuries. How do the electrochemical signals in our brain lead to the rich and varied experiences of thinking, feeling, and being aware? One intriguing way to approach this mystery is by considering our thoughts as quantum states—potentialities that exist in various forms until we focus on them, much like particles in quantum mechanics.

In quantum mechanics, particles can exist in multiple states simultaneously—a concept known as superposition. It is only when these particles are observed that they "collapse" into a definite state. This fascinating principle of quantum physics can be applied to our understanding of thoughts. Just as a particle exists in various potential states, our thoughts may also exist in multiple forms until we bring them into focus through consciousness. This chapter delves into how the principles of quantum mechanics may influence the way we think and how the concept of thoughts as quantum states connects to the broader idea of super-consciousness.

1. Nature of Thoughts

Thoughts are more than just fleeting moments in our minds; they are intricate patterns of brain activity that hold and process information, emotions, and intentions. Understanding thoughts as quantum states allows us to explore the dynamic nature of how we think, make decisions, and interact with the world around us.

Information Encoding

When you study for an exam, your brain works tirelessly to encode all the information—facts, dates, concepts, and methods—that you need to remember. This process is akin to how quantum states encode information about particles in the quantum world. Just as a particle's state contains the information that determines its behavior, your thoughts carry the knowledge that will guide your actions and decisions during the exam.

The brain operates like an advanced computer, where each thought represents a specific data point or quantum state. These states contain the necessary information that allows us to function in our daily lives. For example, when you learn a new skill, your brain encodes the steps required to perform that skill, storing them in neural pathways that can be accessed when needed.

Thoughts also play a crucial role in emotional processing. When you experience a significant event, your brain encodes the emotions associated with that event, creating a memory that influences how you react to similar situations in the future. These emotional memories, much like quantum states, can be recalled and affect your behavior long after the original event has passed.

Dynamic Processes

Thoughts are not static; they are dynamic and constantly evolving based on new information and experiences. This characteristic is similar to how quantum states change when particles interact with their environment. Just as a quantum particle can shift from one state to another due to external influences, our thoughts can change and adapt as we encounter new ideas, experiences, and information.

Consider a situation where you hear a new piece of information during a lecture. Your thoughts immediately adjust to incorporate this new knowledge, demonstrating their flexibility and responsiveness. This dynamic nature of thoughts is what allows us to learn, adapt, and grow. In this way, thoughts can be seen as a complex web of interconnected quantum states, each one influencing and being influenced by the others.

The adaptability of thoughts is crucial for problem-solving and creativity. When faced with a challenge, your brain rapidly shifts through various possible solutions, evaluating each one before settling on the most effective course of action. This process mirrors the behavior of quantum particles, which exist in a state of superposition until they are observed or interact with something else. Your thoughts, too, exist in multiple potential forms until you consciously focus on one, making it your reality.

2. Superposition of Thoughts

One of the most fascinating principles of quantum mechanics is superposition—the idea that particles can exist in multiple states at once until they are observed. Similarly, our thoughts can exist in multiple potential forms until a conscious decision is made, at which point one thought or action is selected. This concept of superposition in the realm of thoughts opens up intriguing possibilities for understanding creative thinking and decision-making processes.

Creative Thinking

Creative thinking often involves holding multiple ideas in superposition, allowing your brain to explore various possibilities simultaneously. When you brainstorm ideas for a project, your brain might hold several potential solutions in superposition, contemplating them all at once. This process allows you to weigh different options and consider various perspectives before arriving at the best solution.

For instance, imagine you are designing a new product. Your brain might simultaneously consider different features, designs, and functions, each representing a different "state" of thought. By allowing these ideas to coexist in superposition, you give yourself the flexibility to combine them in novel ways, leading to innovative and creative solutions.

This superposition of thoughts is not limited to creative pursuits; it is also at play in everyday decision-making. Whether you are deciding what to have for dinner or how to approach a complex problem at work, your brain is constantly juggling multiple options in a state of superposition until you make a conscious decision.

Decision Making

The superposition of thoughts is particularly evident in decision-making processes. Imagine you are considering whether to take a new job. Your brain might evaluate multiple scenarios simultaneously—thinking about the salary, work-life balance, career growth, commute, and other factors. Each of these considerations exists in a state of superposition, allowing your brain to assess all possible outcomes before making a final decision.

This ability to hold and consider multiple outcomes at once is what enables us to make well-rounded, informed decisions. It allows us to weigh the pros and cons of each option and consider how different choices might impact our lives. Once a decision is made, the superposition of thoughts collapses into a single outcome, much like a quantum state collapsing into a definite state upon observation.

This process is not only a testament to the brain's complexity but also highlights the role of consciousness in shaping our reality. By focusing on a particular thought or outcome, we effectively "choose" our reality from the myriad possibilities that exist in the quantum state of our minds.

3. Entanglement of Thoughts

Quantum entanglement is a phenomenon where particles become linked in such a way that the state of one immediately influences the state of another, regardless of the distance between them. This concept can also apply to how our thoughts are interconnected, influencing each other in ways that can shape our perceptions, emotions, and actions.

Mental Associations

Have you ever experienced a particular scent triggering a vivid memory? This is an example of how thoughts can become entangled—one thought or sensory experience triggers another, even if they seem unrelated. For instance, the smell of freshly baked cookies might remind you of your grandmother's house, instantly bringing back memories of childhood visits. This entanglement of thoughts shows how different memories and experiences are linked together in your brain, forming a complex web of associations.

These mental associations are not random; they are shaped by your experiences, emotions, and the context in which they were formed. Over time, these entangled thoughts can influence your behavior and decisions, often in subtle ways. For example, a positive memory associated with a particular song might lead you to choose that song when you need a mood boost, demonstrating how entangled thoughts can guide your actions.

Collective Consciousness

On a larger scale, the concept of entanglement can be applied to collective consciousness—the shared thoughts, beliefs, and intentions of a group of people. Think of a large group of individuals working together towards a common goal, such as a community coming together after a natural disaster. Their collective thoughts and efforts are like entangled particles, connected and influencing each other to achieve a common purpose. This shared experience can create a powerful sense of unity and collective consciousness, where the thoughts and actions of each individual contribute to the overall outcome.

Collective consciousness can also manifest in social movements, where the shared intentions and actions of a group of people can lead to significant changes in society. For example, the collective desire for social justice and equality has fueled many movements throughout history, demonstrating the power of entangled thoughts to bring about real-world change.

This concept of entangled thoughts extends beyond individual and group experiences; it also suggests a deeper connection to the universe itself. Just as entangled particles can influence each other across vast distances, our thoughts may be connected to a larger, universal consciousness—a super-consciousness that links everything in existence.

4. Interaction with Super-Consciousness: Connecting Thoughts with Universal Forces

The concept of super-consciousness suggests a universal force or awareness that connects everything in the universe. This force is not just passive; it actively interacts with everything, including our thoughts and actions. If we consider thoughts as quantum states—tiny, powerful units of potential—they might have the capacity to shape reality, influenced by and influencing super-consciousness.

Influence of Super-Consciousness: The Feedback Loop of Thought and Reality

Super-consciousness can be thought of as a guiding force that maintains balance and harmony in the universe. A useful way to understand this interaction is through the concept of a feedback loop, where our thoughts influence the world around us, and in turn, the world influences our thoughts.

The Feedback Loop: Consider how your mindset can affect the outcome of a situation. For instance, when you approach a challenging situation with positivity—confidence, optimism, and determination—you're likely to take actions that reflect this mindset. Others around you may pick up on your positivity and respond similarly, creating a cycle of positive energy. This feedback loop is a two-way interaction: your thoughts (positive quantum states) influence your actions and environment, and the positive responses from the environment reinforce your thoughts.

This feedback loop is not limited to personal experiences; it can also operate on a larger scale. For example, a community that collectively believes in and works towards a common goal can create a powerful feedback loop that amplifies their efforts and leads to significant achievements. This phenomenon is often seen in successful social movements, where the shared intentions of a group create a momentum that drives change.

In the context of super-consciousness, this feedback loop suggests that our thoughts are not isolated; they are part of a larger, interconnected system that influences and is influenced by the universe. When we align our thoughts with positive intentions, we tap into this broader, universal force, creating outcomes that reflect a harmonious state of being. It's as if our thoughts are in tune with the rhythm of the universe, and super-consciousness amplifies this alignment to produce positive results.

Manifestation and Reality: How Thoughts Shape the World Around Us

The interaction between our thoughts and super-consciousness may not only influence how we perceive the world but also how reality unfolds. This is where the idea that thoughts can influence reality comes into play, often discussed through concepts like the Law of Attraction and Mind-Matter Interaction.

Law of Attraction: The Law of Attraction suggests that like attracts like—meaning that the energy you put out into the universe through your thoughts and emotions can attract similar energy back to you. For example, if you focus on positive thoughts—visualizing a successful outcome, feeling confident, and believing in your ability to succeed—you're more likely to act in ways that lead to success. This positive mindset influences your preparation, delivery, and how others perceive you, all of which contribute to the successful outcome you envisioned.

In the context of super-consciousness, your thoughts (quantum states) might be interacting with this universal force to bring about the desired result. The positive energy you project aligns with the balance and harmony that super-consciousness seeks to maintain, thereby manifesting your intentions into reality. It's as if super-consciousness helps to "co-create" reality with you, responding to the quality and intention of your thoughts.

Mind-Matter Interaction: Another way to understand this interaction is through the concept of mind-matter interaction, where focused thoughts can have a tangible impact on physical reality. Consider a sports team preparing for a game. The players' mental focus, visualization, and collective intention can significantly influence their performance on the field. Their thoughts—focused, intentional quantum states—translate into physical actions that can lead to victory.

This mind-matter interaction shows how consciousness, when concentrated and aligned, can directly influence physical outcomes. It's similar to how observing a quantum system in physics can change its state; in the same way, your focused thoughts can shape the reality around you. Super-consciousness might play a role here by facilitating the alignment between your intentions (quantum states) and the physical world, helping to manifest the desired outcomes.

5. Implications for Personal Development

Understanding thoughts as quantum states underscores the importance of mindfulness and intentional thinking in shaping one's reality and well-being. By learning to harness the power of your thoughts, you can create positive changes in your life and contribute to the collective consciousness.

Mindfulness and Thought Control

Mindfulness is the practice of being fully present and aware of your thoughts, emotions, and surroundings. When you practice mindfulness, you gain greater control over your thoughts, allowing you to guide them towards positive outcomes. For example, when faced with a difficult situation at work, practicing mindfulness can help you stay calm and focused, enabling you to approach the problem with clarity and composure.

Positive thinking is a key aspect of mindfulness. By focusing on positive thoughts and maintaining a constructive attitude, you can create a ripple effect that influences your actions and the outcomes of your efforts. This positive energy contributes to your personal success and overall well-being, reinforcing the idea that your thoughts play a key role in shaping your reality.

Mindfulness Practices: There are various mindfulness practices that can help you cultivate positive thought patterns. Meditation, for instance, is a powerful tool for calming the mind and gaining insight into your thoughts. By regularly practicing meditation, you can develop the ability to observe your thoughts without becoming attached to them, allowing you to choose which thoughts to focus on and which to let go of.

Journaling is another effective mindfulness practice. Writing down your thoughts and emotions can help you process them and gain clarity on your feelings. By reflecting on your thoughts in a journal, you can identify patterns and make conscious choices about how to respond to different situations.

Collective Consciousness and Social Harmony

The idea that our thoughts are interconnected within a larger super-consciousness suggests that collective thinking can have a significant impact on social harmony and global well-being. When people come together with shared intentions, they create a powerful collective consciousness that can drive positive change.

Collective Intentions: Consider a global movement for environmental protection. When millions of people share the same intention to protect the planet, their collective thoughts and actions can lead to real-world changes, such as new environmental policies or conservation efforts. This collective effort is similar to how entangled particles influence each other, showing how collective consciousness can drive positive change.

Shared Experience: After a natural disaster, communities often come together to support each other. The shared experience of working towards recovery creates a strong sense of unity and empathy, much like how entangled thoughts in the brain are connected and influence each other. This interconnectedness can lead to stronger social bonds and a more harmonious society.

Collective consciousness also plays a role in cultural and social movements. Throughout history, groups of people have come together to advocate for change, whether it's for civil rights, gender equality, or environmental sustainability. The shared intentions and actions of these groups create a powerful force that can bring about significant societal changes.

Conclusion

By viewing our thoughts as quantum states, we gain a deeper understanding of how our minds work and how our intentions can influence reality. This perspective allows us to see the profound connection between our thoughts, consciousness, and the super-consciousness that permeates the universe. As we align our thoughts with positive intentions, we tap into the universal force of super-consciousness, co-creating our reality and contributing to the harmony of the universe.

In essence, the principles of quantum mechanics provide a powerful framework for understanding the relationship between our thoughts and the larger reality. By becoming more mindful of our thoughts and intentions, we can harness the power of super-consciousness to shape our lives in meaningful and fulfilling ways. This chapter lays the foundation for exploring how these concepts can be applied in everyday life, guiding us towards greater personal growth, well-being, and a deeper connection with the universe.

Story Time: Ananya's Struggle with Overwhelm and Stress

Ananya was a hardworking project manager at a big tech company. She was known for juggling multiple projects at once and doing her job well. But as her workload grew, Ananya began to feel overwhelmed. The constant pressure to meet deadlines and always be available started to affect her mental and physical health. Ananya believed that stress was just part of the job and that managing it was impossible. She thought, "Only people with fewer responsibilities or more support can handle their work without falling apart."

Problem: Chronic Stress Due to Overwhelm

Ananya's inability to manage her workload and the resulting stress led to burnout, poor sleep, and strained relationships. She felt stuck, thinking that stress was an unavoidable part of her career and that managing it was beyond her control.

Solution: Understanding and Harnessing Thoughts as Quantum States

Realizing that her stress was getting out of control, Ananya decided to seek help. She attended a seminar on mindfulness and stress management, where the speaker introduced the idea of thoughts as quantum states. The idea was that our thoughts exist in a state of potential, and by focusing our mind, we can influence which potential state becomes reality. This concept intrigued Ananya, who began to see her stress as something she could control by changing her thoughts.

Awareness of Stress-Inducing Beliefs and Mindset Shift

Ananya started by identifying the limiting beliefs that were causing her stress. She realized that her belief in stress as an unavoidable part of her job was holding her back from finding solutions. By recognizing this belief as just one possible thought among many, she opened herself to the idea that she could manage her stress more effectively.

Identifying Emotional Triggers and Thought Patterns

Through mindfulness practices, Ananya became more aware of the specific thoughts and emotions that triggered her stress. She noticed that she often felt overwhelmed when she thought about all her responsibilities at once. By breaking down her workload into smaller, more manageable tasks and focusing on one thing at a time, she was able to reduce her feelings of overwhelm.

Reframing Thoughts and Visualization

Ananya began to reframe her thoughts about her workload. Instead of thinking, "I'll never get everything done," she started telling herself, "I can handle one task at a time." She also used visualization techniques to imagine herself successfully managing her workload, feeling calm and in control. This positive visualization helped shift her mindset from one of stress and anxiety to one of confidence and calm.

Implementing Practical Stress Management Techniques

With a new mindset, Ananya put practical strategies in place to manage her workload. She started by prioritizing her tasks based on importance and urgency, making sure she focused on the most important work first. She also began delegating tasks to her team members more effectively, trusting them to handle responsibilities without her constant oversight.

Setting Boundaries and Self-Care

Ananya realized that managing stress also meant setting boundaries around her work. She decided to establish clear work hours and made an effort to disconnect from work during her personal time. She also added self-care practices to her daily routine, like regular exercise, meditation, and spending time with loved ones. These activities helped her recharge and maintain her well-being.

Embracing Mindfulness and Staying Present

Ananya continued practicing mindfulness to stay present and focused throughout the day. By becoming more aware of her thoughts and how they impacted her stress levels, she was able to catch herself before spiraling into overwhelm and redirect her focus to what she could control in the moment.

Over time, Ananya noticed a significant reduction in her stress levels. She was better able to manage her workload, felt more in control of her time, and experienced improved overall well-being. Ananya realized that by understanding and harnessing the quantum nature of her thoughts, she could change her reality and effectively manage her stress.

Key Insight:

The idea of thoughts as quantum states suggests that our mindset and focus can shape our reality, including how we experience and manage stress.

By becoming aware of limiting beliefs, reframing negative thoughts, and using practical stress management techniques, we can reduce overwhelm and improve our well-being. Ananya's story shows how understanding and applying these principles can help overcome chronic stress and lead to a more balanced and fulfilling life.

CHAPTER 12

Consciousness and Quantum Interaction

Introduction

"Consciousness is the key to unlocking the mysteries of the universe."

– Max Planck

The concept of consciousness has intrigued humans for centuries, spanning the fields of philosophy, psychology, and, more recently, quantum mechanics. As we delve deeper into the nature of reality, it becomes increasingly evident that consciousness plays a pivotal role in shaping our experiences and interactions with the world around us. But what if consciousness is not just a passive observer but an active participant in the quantum realm? This chapter explores the profound connection between consciousness and quantum processes, examining how our awareness might influence the very fabric of reality.

In the previous chapter, we explored how thoughts might function as quantum states—potentialities that exist until we bring them into focus through consciousness. Now, we will take this concept further by examining the dynamic interaction between consciousness itself and the quantum processes that underlie reality. Could our consciousness be more than just a byproduct of brain activity? Could it be an integral part of a larger, universal consciousness that governs the universe?

1. The Nature of Thoughts and Consciousness

To understand how consciousness interacts with quantum processes, it is essential to first clarify what we mean by "thoughts" and "consciousness." These two concepts, while closely related, have distinct characteristics and functions.

What Are Thoughts?

Thoughts are the building blocks of our mental landscape. They encompass a wide range of cognitive processes, including ideas, memories, emotions, and intentions. Thoughts can be fleeting or persistent, simple or complex, and they play a crucial role in shaping our perception of reality.

Different Kinds of Thoughts: Thoughts can be categorized into various types based on their origin and function. Sensory thoughts are linked to our immediate experiences, such as the sight of a beautiful sunset or the smell of fresh coffee. Memories, on the other hand, are thoughts that connect us to past experiences, allowing us to recall and reflect on events that have shaped our lives. Abstract thoughts involve higher-order thinking, such as problem-solving, planning, and creativity. Each type of thought serves a unique purpose in helping us navigate the world and make sense of our experiences.

How Thoughts Work: The brain processes thoughts through a complex network of neurons that communicate via electrical and chemical signals. These neurons form intricate patterns of activity that correspond to different types of thoughts. For example, when you recall a memory, specific neurons associated with that memory become active, allowing you to bring the past experience into your present awareness. Similarly, when you solve a problem, your brain engages in a series of thought processes that involve evaluating options, predicting outcomes, and making decisions. This dynamic interplay of neuronal activity is what allows us to think, learn, and adapt to new situations.

What Is Consciousness?

Consciousness is the state of being aware of oneself and one's surroundings. It is the experience of being alive, perceiving the world, and having thoughts and emotions. Consciousness is not just about being awake; it encompasses the entire spectrum of awareness, from full alertness to deep introspection.

The Many Layers of Consciousness: Consciousness is often described as having multiple layers or levels. At the most basic level, there is the consciousness of being awake and aware of the external world—what we might call "ordinary" consciousness. However, there are also deeper layers of consciousness that involve self-reflection, meditation, and altered states of awareness. For example, during meditation, one might experience a heightened sense of awareness that transcends ordinary consciousness, allowing for deeper insights and a sense of connection to something greater than oneself.

Self-Awareness: The Mirror of the Mind: A key aspect of consciousness is self-awareness—the ability to reflect on one's own thoughts, emotions, and experiences. Self-awareness allows us to recognize ourselves as distinct individuals with our own unique perspectives and experiences. It is like looking into a mirror, not just to see our physical reflection but to understand our inner selves—our motivations, desires, and fears. This self-awareness is what makes us conscious beings, capable of introspection and personal growth.

2. Consciousness in the Context of Quantum Mechanics

Quantum mechanics, the branch of physics that deals with the behavior of subatomic particles, has introduced concepts that challenge our traditional understanding of reality. Two key ideas—quantum states and the observer effect—are particularly relevant to the study of consciousness.

Thoughts as Quantum States

In the quantum world, particles can exist in multiple states simultaneously, a phenomenon known as superposition. Similarly, thoughts can be seen as potentialities that exist in a superposed state until they are brought into focus through consciousness. Each thought represents a possible state of mind, and it is our conscious attention that "collapses" these potentialities into a single, experienced reality.

When you consider a decision, for example, your mind may entertain multiple possibilities simultaneously—each possibility representing a different quantum state of thought. It is only when you consciously choose one option that this superposition collapses, and the chosen thought becomes your reality. This process highlights the role of consciousness as an active participant in shaping our mental landscape.

Consciousness as the Observer Effect

In quantum mechanics, the observer effect refers to the idea that the act of observation can influence the outcome of a quantum event. This concept can be extended to consciousness, where our awareness plays a crucial role in determining the state of our thoughts and, by extension, our reality.

Consciousness acts as the observer that brings thoughts into focus, collapsing their potentialities into specific experiences. Without consciousness, thoughts would remain in a state of superposition, existing as mere possibilities rather than actualized realities. This observer effect underscores the power of consciousness in shaping our perceptions, decisions, and interactions with the world.

For instance, when you observe your thoughts during meditation, you may notice that simply being aware of a thought can change its nature. A negative thought, once observed, may lose its intensity or even transform into a more positive one. This illustrates how consciousness, like the observer in quantum mechanics, has the power to alter the state of our mental processes.

3. The Dynamic Interaction Between Thoughts and Consciousness

The relationship between thoughts and consciousness is not static; it is a dynamic, ongoing interaction that shapes our experiences and influences our reality. Understanding this interaction can provide valuable insights into how we can harness the power of our minds to create positive change in our lives.

Conscious Thought Processes: The Power of Awareness

Consciousness provides the platform upon which our thoughts play out, enabling us to engage in complex cognitive activities such as reasoning, introspection, and creative thinking. By directing our conscious attention to specific thoughts, we can enhance our cognitive abilities and make more informed decisions.

Attention and Focus: Consciousness allows us to focus on specific thoughts while filtering out distractions. This selective attention is akin to tuning a radio to a particular station, allowing us to concentrate on the information that is most relevant to our current task. For example, when studying for an exam, you can consciously direct your attention to the material at hand, blocking out irrelevant thoughts and distractions. This focused attention not only improves your ability to retain information but also enhances your overall cognitive performance.

Reflective Thinking: Consciousness also enables reflective thinking—the ability to step back and examine your thoughts, emotions, and experiences. This form of thinking allows you to analyze your reasoning, question your beliefs, and make deliberate decisions based on careful consideration. Reflective thinking is crucial for personal growth, as it helps you identify patterns in your behavior, recognize areas for improvement, and develop a deeper understanding of yourself and the world around you.

Unconscious Influences: The Hidden Drivers

While conscious thought often takes center stage, much of our thinking is influenced by processes occurring in the unconscious mind. These hidden drivers play a significant role in shaping our thoughts, emotions, and behaviors, often without our conscious awareness.

Priming: Priming occurs when exposure to a stimulus influences your response to a subsequent stimulus, often without your conscious awareness. For example, hearing a particular song might unconsciously prime you to feel nostalgic, bringing up memories associated with that song. This priming effect demonstrates how unconscious processes can shape your thoughts and emotions, even when you are not actively aware of them.

Emotional Regulation: The unconscious mind also plays a crucial role in managing your emotional responses. For instance, you might feel anxious without fully understanding why, only to realize later that your anxiety was triggered by an unconscious association with a past experience. This unconscious regulation of emotions can influence your conscious thoughts and decisions, often steering your behavior in subtle ways.

The Feedback Loop: Thoughts Shaping Consciousness, and Vice Versa

The interaction between thoughts and consciousness is a continuous feedback loop, where each influences the other in a dynamic interplay that shapes your mental landscape.

Self-Reflection: Engaging in self-reflection allows you to consciously examine your thoughts, leading to greater self-awareness and personal growth. By reflecting on your thoughts, you can challenge old beliefs, adopt new perspectives, and alter your behavior in positive ways. This process of self-reflection creates a feedback loop, where conscious awareness influences your thoughts, and your thoughts, in turn, shape your consciousness.

Mindfulness: Mindfulness is the practice of being fully present and aware of your thoughts, emotions, and surroundings. By practicing mindfulness, you can fine-tune the feedback loop between thoughts and consciousness, allowing you to manage your mental state more effectively. For example, by becoming aware of negative thought patterns, you can consciously choose to redirect your thoughts toward more positive and constructive ideas. This practice helps you maintain a balanced mind and enhances your overall well-being.

4. Super-Consciousness and Thought Interaction

The concept of super-consciousness suggests that there is a universal force or intelligence that subtly shapes and influences our individual thoughts, guiding our thinking in ways that align with larger, universal patterns. This interaction between super-consciousness and our thoughts opens up new possibilities for understanding the nature of reality and our place within it.

Influence of Super-Consciousness on Individual Thoughts: The Bigger Picture

Super-consciousness can be thought of as a guiding force that maintains balance and harmony in the universe. It is akin to a universal mind that interacts with individual consciousness, shaping our thoughts and actions in alignment with a greater cosmic order.

Collective Unconscious: This idea resonates with Carl Jung's concept of the collective unconscious—a shared reservoir of experiences and archetypes common to all humans. Just as the collective unconscious influences our dreams, behaviors, and cultural expressions, super-consciousness might influence our thoughts, helping us tap into universal wisdom and insights. This connection to a larger consciousness suggests that our thoughts are not isolated but are part of a broader, interconnected web of consciousness that spans the entire universe.

Individual Contributions to Super-Consciousness: A Two-Way Street

While super-consciousness may influence our thoughts, our thoughts and intentions can also contribute to the state of super-consciousness. This reciprocal relationship suggests that we are not merely passive recipients of cosmic influence but active participants in the ongoing creation of reality.

Energetic Imprints: Every thought and intention we hold creates an energetic imprint that interacts with the universal web of consciousness. These imprints contribute to the overall state of super-consciousness, suggesting that our mental activities have a broader impact on the universe. For example, when you focus on positive thoughts and intentions, you contribute to the positive energy of the collective consciousness, potentially influencing outcomes on a larger scale. This idea reinforces the notion that we are co-creators of reality, working in tandem with super-consciousness to shape the world around us.

5. Practical Implications of Thought and Consciousness Interaction

Understanding the interaction between thoughts and consciousness, particularly in the context of super-consciousness, has profound implications for personal development, mental well-being, and ethical decision-making.

Enhancing Cognitive Flexibility: Adapting to Life's Challenges

Cognitive flexibility is the ability to adapt your thinking to new situations and challenges. By understanding how thoughts and consciousness interact, you can enhance your cognitive flexibility and become more resilient in the face of change.

Mindfulness Practices: Mindfulness meditation is an effective way to enhance cognitive flexibility. By regularly practicing mindfulness, you can increase your awareness of your thoughts and emotions, allowing you to adapt more easily to new circumstances. For example, if you encounter an unexpected challenge at work, mindfulness can help you stay calm and focused, enabling you to approach the situation with a clear and open mind.

Improving Mental Well-Being: Fostering Positive Patterns

Recognizing the feedback loop between thoughts and consciousness can significantly impact mental well-being. By cultivating positive thought patterns and harnessing the power of consciousness, individuals can foster a healthier, more balanced state of mind.

Positive Thinking: Positive thinking involves consciously focusing on thoughts that promote well-being and resilience. When you maintain positive thoughts, you create a feedback loop that reinforces optimistic and constructive attitudes, which can lead to improved mental health. For instance, by deliberately practicing gratitude, you can shift your focus from what is lacking in your life to what you appreciate, thereby enhancing your overall sense of happiness and satisfaction. This shift in focus not only improves your mood but also influences your interactions with others, creating a ripple effect of positivity.

Mindfulness Practices: Practicing mindfulness can help break negative thought patterns and replace them with healthier ones. When you become aware of negative or self-defeating thoughts, mindfulness allows you to acknowledge them without judgment and gently guide your mind toward more positive or neutral thoughts. For example, if you notice yourself engaging in negative self-talk, mindfulness can help you recognize this pattern and consciously choose to replace it with more encouraging and supportive thoughts. Over time, this practice can lead to a more balanced and peaceful state of mind.

Ethical and Moral Development: Guiding Our Actions

The interaction between thoughts and consciousness plays a significant role in our ethical and moral development. By reflecting on our thoughts and aligning them with our values, we can make more ethical decisions and contribute to a more just and harmonious society.

Reflective Practices: Reflective practices, such as journaling or meditation, provide opportunities for self-examination and moral reasoning. By regularly engaging in these practices, you can develop a deeper understanding of your values and how they influence your decisions. For example, by reflecting on a difficult decision, you can consider the ethical implications of your choices and align your actions with your principles. This process of reflection helps to cultivate a strong moral compass and promotes ethical behavior in daily life.

Conscious Decision-Making: Consciousness allows us to deliberate on our actions and their potential impact on others. By bringing awareness to our decisions, we can ensure that they are guided by empathy, compassion, and a commitment to doing what is right. For instance, when faced with a moral dilemma, consciously considering the perspectives and well-being of all involved parties can lead to more just and fair outcomes. This conscious decision-making process is essential for fostering ethical behavior and contributing to the greater good.

6. Consciousness and Quantum Interaction in Daily Life

The concepts of consciousness and quantum interaction are not just theoretical; they have practical applications that can enhance your daily life. By understanding and applying these principles, you can harness the power of your mind to create positive change and achieve your goals.

Practice Selective Focus to Shape Reality

Selective focus involves consciously directing your attention to specific thoughts or goals, which can help you shape your reality and bring your desires to fruition.

Application: Start your day by setting a clear intention or goal to focus on throughout the day. For example, if your goal is to improve your productivity at work, consciously direct your attention to tasks that align with this goal. By maintaining focus on your desired outcome, you increase the likelihood of achieving it. This practice is similar to the observer effect in quantum mechanics, where focused observation influences the outcome of a quantum system. In daily life, selective focus acts as a powerful tool for manifesting your intentions.

Use Thought Reflection to Enhance Decision-Making

Reflecting on your thoughts can improve your decision-making process by allowing you to consider multiple perspectives and outcomes before making a choice.

Application: Before making a significant decision, take a moment to reflect on the possible outcomes. Consider how each option aligns with your values, goals, and the potential impact on others. For example, if you are deciding whether to accept a new job offer, reflect on how the position aligns with your career aspirations, work-life balance, and long-term goals. This reflective process helps you make more informed and thoughtful decisions, leading to better outcomes.

Cultivate Mental Awareness to Identify Cognitive Patterns

Mental awareness involves being attuned to the thoughts that frequently occupy your mind, which can help you identify patterns that may be limiting or unproductive.

Application: Keep a thought journal to track recurring thoughts and identify patterns that may be influencing your behavior and decisions. For example, if you notice that you frequently think about past failures, this awareness can help you recognize a pattern of self-doubt. Once identified, you can consciously work to challenge and reframe these thoughts, replacing them with more empowering and constructive beliefs. This practice not only enhances self-awareness but also promotes personal growth and resilience.

Harness the Power of Mental Associations

Mental associations involve linking positive thoughts with challenging situations, which can help you approach difficulties with greater confidence and creativity.

Application: When faced with a stressful situation, recall a past success or positive experience to boost your confidence and motivation. For example, if you are preparing for a challenging presentation, mentally revisit a time when you successfully delivered a presentation in the past. This positive association can help reduce anxiety and enhance your performance. By consciously creating and reinforcing positive mental associations, you can approach challenges with a more optimistic and empowered mindset.

Engage in Creative Visualization for Problem-Solving

Creative visualization is a powerful tool that allows you to explore different solutions to a problem before choosing the best course of action.

Application: Visualize various solutions to a problem and consider the potential outcomes of each option. For example, if you are planning a project, visualize the steps involved in completing it, the challenges you might face, and the successful outcome. This mental rehearsal helps you anticipate obstacles and develop effective strategies for overcoming them. Creative visualization not only enhances problem-solving skills but also boosts confidence and motivation by allowing you to mentally rehearse success.

7. Connecting Thought with Consciousness and Consciousness with Super-Consciousness

The ultimate goal of understanding the interaction between thoughts and consciousness is to connect with the larger, universal consciousness—what we refer to as super-consciousness. By bridging thought with consciousness and merging consciousness with super-consciousness, you can tap into the profound wisdom and creative power that underlies the universe.

Bridging Thought and Consciousness

The first step in this process is to practice intentional reflection, which involves consciously engaging with your thoughts to bring them into clear focus.

Practice: Each day, take a few moments to consciously reflect on your thoughts. Write down your thoughts in a journal, focusing on those that are most significant or recurring. This simple act of writing forces you to slow down and engage your consciousness with each thought, making it more deliberate and focused. Just as in quantum mechanics where observing a particle defines its state, consciously reflecting on your thoughts solidifies and clarifies them. Over time, this practice helps to strengthen the connection between your thoughts and your conscious awareness.

Merging Consciousness with Super-Consciousness

To connect your individual consciousness with the universal super-consciousness, engage in activities that expand your awareness beyond yourself and your immediate surroundings.

Practice: Engage in meditation, spend time in nature, or practice empathy by deeply listening to others. These activities help you step out of your personal bubble and connect with something larger. For example, during meditation, focus on the idea that your consciousness is part of a greater whole, much like how individual particles are part of a larger wave in quantum mechanics. This practice helps you feel more in tune with the universal flow of consciousness and opens you to insights and wisdom that transcend ordinary awareness.

Regular Practice and Integration

The final step is to integrate these practices into your daily routine, making them a natural part of your life.

Practice: Start your day with a brief reflection on your thoughts and end it with a few minutes of meditation or quiet time. Over time, this regular practice will help you naturally bridge your thoughts with your consciousness and align your consciousness with the larger super-consciousness. It's about consistently making small, conscious efforts to be aware, reflect, and connect, which gradually leads to a deeper sense of unity with the world around you. As you continue this practice, you may find that your thoughts and actions become more aligned with the universal consciousness, leading to a more harmonious and fulfilling life.

Conclusion

In this chapter, we explored the profound connection between consciousness and quantum interaction, highlighting how our thoughts, when aligned with our consciousness, can influence and shape reality. By understanding and applying the principles of quantum mechanics to our mental processes, we can enhance our cognitive abilities, improve our mental well-being, and connect with the larger super-consciousness that governs the universe. The journey of connecting thought with consciousness and consciousness with super-consciousness is an ongoing process that requires intention, awareness, and practice. As you continue to explore this connection, you may discover new insights and possibilities that enrich your life and bring you closer to the essence of the universe.

Story Time: Vikram's Struggle with Time Management

Vikram was a skilled software developer working at a busy tech company. Even though he was good at his job, he often felt overwhelmed by all the tasks he had to manage. Deadlines always seemed impossible to meet, and he often stayed late at the office, only to come home feeling tired and stressed. Vikram believed that managing time well was something other people could do because they had more resources, better education, or fewer responsibilities. He thought, "I'm not a machine; I'm just a human. Other people manage their time because they have advantages I don't."

Problem: Time Management Issues

Vikram's struggle with managing his time led to constant stress, missed deadlines, and a poor balance between work and personal life. He was convinced that effective time management was something only people with better resources could achieve, which stopped him from looking for solutions.

Solution: Understanding and Using the Power of Thoughts

As Vikram's stress began to affect his health and personal life, he knew something had to change. He attended a productivity workshop where the speaker explained that our thoughts, like particles in the quantum world, have the potential to shape our reality. By focusing on positive and organized thoughts, we can influence our reality and manage time better.

Recognizing Limiting Beliefs and Changing His Mindset:

Vikram started by noticing the negative beliefs that were holding him back. He realized that thinking time management was impossible for him was just a thought, not a fact. By acknowledging this, he opened himself up to the possibility that he could improve his time management skills.

Understanding Emotional Triggers:

Through reflection and mindfulness, Vikram identified that his stress about time was often triggered by a fear of failure and a need to prove himself. This fear led him to take on too much work, which made him feel overwhelmed. By addressing these emotional triggers, Vikram was able to reduce the pressure he placed on himself and focus on what really mattered.

Visualizing Success and Changing His Thoughts:

Vikram began each day by visualizing how he wanted his day to go. Instead of seeing his tasks as a mountain he couldn't climb, he imagined himself managing his time well, completing his tasks easily, and still having time to relax. This positive visualization helped him shift from thinking, "I don't have enough time," to believing, "I can manage my time effectively."

Using Practical Time Management Strategies:

Vikram started using practical time management techniques that matched his new mindset. He prioritized his tasks based on urgency and importance, broke down large projects into smaller tasks, and used time-blocking to stay focused during specific periods of the day.

Delegating Tasks and Setting Boundaries:

Vikram realized that managing time well also means knowing your limits. He began delegating tasks to his team and setting clear boundaries at work. He communicated better with his colleagues, set realistic deadlines, and made sure to stop working at a certain time each day, which helped him avoid burnout.

Practicing Mindfulness and Staying Focused:

Vikram added mindfulness practices to his daily routine to help him stay focused and present. By being more aware of his thoughts and how they influenced his actions, he was able to stay on task and avoid the distractions that had previously slowed him down.

Over time, Vikram noticed a big improvement in his ability to manage time. He was meeting deadlines more consistently, feeling less overwhelmed, and enjoying a better balance between work and life. Vikram realized that by understanding the power of his thoughts and making conscious, practical changes, he could transform his approach to time management.

Key Insight:

The idea that our thoughts have the power to shape our reality suggests that our mindset and focus can greatly influence how we manage our time. By becoming aware of negative beliefs, addressing emotional triggers, and using practical time management strategies, anyone can overcome challenges related to productivity and balance. Vikram's story shows how understanding the power of focused thoughts and making intentional changes can lead to better time management and reduced stress.

CHAPTER 13

Quantum Tunneling and Decoherence

Introduction

"The more you find out about the world, the more opportunities there are to laugh at it."

– Bill Nye

As we dive into the mysterious world of quantum mechanics, we come across ideas that seem almost magical. Two of these ideas are quantum tunneling, where particles move through barriers they shouldn't be able to cross, and quantum decoherence, which explains how quantum systems lose their unique properties as they interact with their surroundings. These concepts challenge our understanding of reality and offer new ways to think about consciousness and the idea of super-consciousness—a universal force that connects and harmonizes everything in existence. In this chapter, we'll explore how quantum tunneling and decoherence can help us understand super-consciousness better.

Quantum Tunneling: Crossing the Barriers of Reality

1. **The Basics of Quantum Tunneling** Quantum tunneling is when particles pass through energy barriers that, according to classical physics, they shouldn't be able to cross. Imagine standing in front of a wall taller than you can jump. Normally, you couldn't get to the other side without help. But in the quantum world, particles like electrons can "tunnel" through these walls, appearing on the other side as if the barrier wasn't there.

   Example: It's like a magician passing through a solid wall—something that seems impossible in the real world, but particles in the quantum world do it regularly. This tunneling isn't just theoretical; it happens in real life, like in semiconductors, which are crucial for all modern electronics. Without quantum tunneling, our computers and smartphones wouldn't work the way they do.

2. **Quantum Tunneling and Super-Consciousness** Now, let's connect this idea to super-consciousness. If we think of consciousness as something that goes beyond the physical limits of the brain, quantum tunneling offers an interesting comparison. Just as particles can move through barriers in the quantum world, our thoughts and consciousness might "tunnel" through the barriers of physical reality. This could explain moments of intuition or sudden insights—when new ideas seem to appear out of nowhere, bypassing our usual logical thinking.

Analogy: Think about a time when you were stuck on a problem, and the solution suddenly came to you when you weren't even thinking about it. This "tunneling" of thought might be similar to how particles in quantum mechanics move through barriers, suggesting that our consciousness might work on a quantum level, finding pathways that aren't visible through normal thinking.

3. **Super-Consciousness as a Guiding Force** If we take this idea further, super-consciousness might be the guiding force that allows these "quantum leaps" in thought. Just as quantum tunneling helps particles move through barriers, super-consciousness might help us go beyond the limits of ordinary thinking, connecting us to deeper insights and a broader understanding of the universe.

Quantum Decoherence: The Transition from Quantum to Classical

1. **Understanding Quantum Decoherence** Quantum decoherence is the process where a quantum system loses its unique quantum properties—like being in multiple states at once—when it interacts with its environment. This interaction causes the system to "decohere," meaning it starts to behave more like everyday objects we're familiar with.

Example: Imagine a spinning top that can spin in multiple directions at the same time (a superposition). When you observe it or it interacts with something else, it suddenly "chooses" a single direction to spin in—that's decoherence. Decoherence helps explain why we don't see quantum effects in our daily lives. Although particles at the quantum level can exist in multiple states at once, these effects disappear when they interact with the larger, classical world.

2. **The Role of Decoherence in Consciousness** Decoherence might also play a role in how consciousness emerges from the brain's quantum processes. The brain, filled with billions of neurons and trillions of connections, could be seen as a complex quantum system. As quantum states within the brain decohere, they might give rise to the stable, consistent experience of consciousness that we all have.

   Analogy: Think of your consciousness as the "final product" of countless quantum processes in the brain. As these processes interact and decohere, they form the clear, singular experience of being awake, aware, and alive.

3. **Super-Consciousness and the Collapse of Possibilities** In the context of super-consciousness, quantum decoherence could be seen as the process by which potential realities are "collapsed" into a single, coherent experience. Just as quantum systems decohere to form stable classical states, super-consciousness might help shape the potential states of the universe into the reality we experience.

   Example: Imagine you're at a crossroads in life, faced with multiple possible choices. The act of making a decision could be seen as a form of decoherence, where super-consciousness helps collapse the potential outcomes into a single path that you then follow.

Connecting Quantum Tunneling and Decoherence to Super-Consciousness

1. **A Unified Understanding** When we bring together the ideas of quantum tunneling and decoherence, we start to see a unified picture of how super-consciousness might work. Quantum tunneling shows how consciousness might transcend barriers and limitations, while quantum decoherence explains how these quantum potentials collapse into the stable, coherent experiences that form our reality.

2. **The Pathways of Consciousness** Just as particles tunnel through barriers and decohere into classical states, our thoughts and consciousness might navigate through the complexities of the universe, guided by super-consciousness. This suggests that super-consciousness isn't just a passive backdrop, but an active force that shapes the flow of reality, helping to create the coherent, meaningful experiences that define our lives.

Practical Implications

1. **Transcending Mental Barriers** Understanding quantum tunneling can inspire us to go beyond the mental barriers that limit our thinking. Just as particles can pass through physical barriers, we can learn to overcome psychological and emotional barriers using the power of focused thought and super-consciousness.

2. **Embracing Change and Uncertainty** Quantum decoherence teaches us about the transition from potential to reality. In our lives, this can be a metaphor for embracing change and uncertainty. Just as quantum states collapse into a single outcome, we can learn to navigate life's uncertainties with the understanding that our choices, guided by super-consciousness, help shape the reality we experience.

Connecting the Dots: A Unified Picture of Super-Consciousness and Quantum Reality

In the past several chapters, we've explored different quantum ideas and how they connect to consciousness, thoughts, and the broader concept of super-consciousness. Now, let's bring these ideas together to form a clear picture that deepens our understanding and shows how these concepts are linked to shape the reality we experience.

1. **The Presence of Consciousness in Everything** In Chapter 8, we discussed the idea that consciousness might be present everywhere in the universe, not just within living beings. We explored how consciousness could be a fundamental force that keeps everything in balance, connecting every particle, energy field, and living organism in a universal web.

2. **Super-Consciousness and the Equilibrium of Strings** Chapter 9 explored how super-consciousness acts like the conductor of the universe, making sure that everything—from the smallest particle to the largest galaxy—vibrates in harmony. This idea was connected to string theory, suggesting that super-consciousness is responsible for maintaining the balance of the strings that make up the fabric of reality.

3. **Karma and Super-Consciousness** In Chapter 10, we introduced the concept of karma within the framework of super-consciousness. We explored how our actions create ripples in the universal web, influencing the balance of energy and matter. Super-consciousness, as a living force, helps restore balance by adjusting the outcomes of actions, making sure that karma is part of the cosmic order.

4. **Quantum States and Thought Processes** Chapter 11 explored the idea that thoughts could be seen as quantum states. Just like quantum particles that exist in multiple states until observed, our thoughts are fluid and full of potential until we focus on them. This concept suggested that our thoughts are powerful forces that can influence reality based on how we "observe" or focus on them.

5. **Consciousness and Quantum Interaction** In Chapter 12, we looked at how consciousness interacts with thoughts. Consciousness acts as the observer that decides which potential realities of our thoughts become real. This interaction isn't one-sided—unconscious processes can also influence how consciousness interacts with thoughts, much like unseen forces can influence quantum systems.

6. **Quantum Tunneling and Crossing Barriers** Chapter 13 introduced quantum tunneling, where particles pass through barriers that seem impossible to cross. This idea extended to our consciousness, suggesting that our minds might also have the ability to "tunnel" through barriers, helping us overcome limitations and achieve insights that seem impossible.

7. **Quantum Decoherence: Shaping Reality** Along with quantum tunneling, we also explored quantum decoherence, which explains how quantum possibilities collapse into a single, stable reality. This

concept helps us understand how super-consciousness might guide the countless potential outcomes of our thoughts into the stable reality we experience.

8. **The Role of Super-Consciousness** Throughout these chapters, we've discussed super-consciousness as the force that connects and harmonizes everything. Super-consciousness isn't just a passive observer but an active participant in the universe, helping to shape the reality we experience by guiding the interaction between quantum states, thoughts, and consciousness.

Key Takeaway: Super-consciousness acts as the guiding force that shapes our thoughts, consciousness, and reality, making sure everything works together in a balanced and harmonious way.

The Unified Picture: How It All Connects

Imagine the universe as a vast, interconnected web where every thought, particle, and action is linked by the threads of quantum mechanics and super-consciousness.

Thoughts as Quantum States: Our thoughts exist in a state of potential, much like quantum particles. They can influence reality based on how we focus and observe them.

Consciousness as the Observer: Consciousness interacts with these thoughts, turning their potential into specific realities. This process is influenced by both conscious and unconscious factors, just like quantum systems are influenced by external forces.

Quantum Tunneling and Overcoming Barriers: Just as particles can tunnel through barriers, our consciousness can transcend limitations, allowing us to achieve breakthroughs and insights that seemed impossible.

Quantum Decoherence and Shaping Reality: Quantum decoherence helps explain how these potential thoughts and actions become the stable, coherent reality we experience. Super-consciousness guides this process, ensuring that the outcome is balanced and meaningful.

Applying the Unified Picture in Daily Life

1. **Harnessing Thought Power:** Recognize that your thoughts have the potential to shape your reality. Focus on positive, constructive thoughts to influence the outcomes you desire.

2. **Mindful Observation:** Be aware of how your consciousness interacts with your thoughts. Practice mindfulness to ensure that you're consciously guiding your thoughts in a direction that benefits your life.

3. **Overcoming Limitations:** Use the concept of quantum tunneling as inspiration to break through mental and emotional barriers. Believe that it's possible to achieve breakthroughs in areas where you previously felt stuck.

4. **Shaping Your Reality:** Understand that your thoughts and actions contribute to the reality you experience. By aligning your intentions with positive outcomes, you can shape a reality that is harmonious and fulfilling.

5. **Connecting with Super-Consciousness:** Embrace the idea that you're part of a larger, interconnected system. Your thoughts and actions are not isolated—they influence and are influenced by the broader universe. Use this connection to guide your decisions and actions in a way that contributes to the greater good.

Conclusion

By exploring concepts like quantum tunneling and decoherence, we gain deeper insights into the nature of super-consciousness and its role in shaping reality. These quantum ideas offer powerful metaphors for understanding how consciousness can break through limitations and how potential realities become the experiences we live. As we continue to explore these ideas, we open up new possibilities for understanding the mind, consciousness, and the universe as a living, interconnected whole.

Story Time: Finding Balance in Work and Life with Quantum Tunneling and Decoherence

Neha was a dedicated marketing manager at a fast-growing startup. She loved her job and worked hard, often taking on more tasks than she could handle.

This led to long hours, constant stress, and a blurry line between her work and personal life. Neha felt overwhelmed and trapped, convinced that achieving work-life balance was impossible. She thought, "Only people with fewer responsibilities or more support can manage work and life without falling apart."

The Problem: Work-Life Imbalance

Neha's struggle to balance her demanding job with her personal life caused her to feel burnt out, damaged her relationships, and hurt her health. She felt stuck in a cycle where work took up all her energy and time, leaving no room for anything else. Her belief that balance was impossible only made her feel more frustrated and hopeless.

The Solution: Using Quantum Ideas to Find Balance

As Neha's stress began to seriously affect her life, she knew she needed to make a change. She attended a wellness seminar where she learned about quantum tunneling and decoherence—two ideas from quantum mechanics. These concepts offered a fresh perspective on breaking through barriers and creating a balanced life. They resonated with Neha, inspiring her to make positive changes.

1. Quantum Tunneling: Breaking Through Mental Barriers

At the seminar, Neha learned about quantum tunneling, where particles can pass through barriers that should be impossible to cross. This made her realize that the barriers she faced in finding work-life balance weren't as solid as they seemed. Just as particles tunnel through physical obstacles, Neha understood that she could "tunnel" through her mental and emotional barriers—those limiting beliefs that told her balance was out of reach.

Simple Explanation: Quantum tunneling happens because particles don't always follow the usual rules—they can sometimes "sneak" through barriers that seem impossible. Neha applied this idea to her life by realizing that the barriers she faced weren't as strong as she thought. She started to believe that balance was possible and began to break through her doubts and fears, creating a path to a new reality where she could manage both work and life.

2. Quantum Decoherence: Creating a Balanced Reality

After understanding how quantum tunneling could help her overcome mental barriers, Neha learned about quantum decoherence. This idea explains how multiple possibilities collapse into a single, clear reality. Neha saw that her chaotic workload was like a system with too many tasks competing for her attention at once. By applying the concept of decoherence, Neha started to "collapse" her tasks into a clear plan, making her responsibilities more organized and manageable.

Simple Explanation: In quantum mechanics, decoherence is when multiple possibilities combine into one clear outcome. Neha used this idea to organize her life. She prioritized her tasks and created a structured plan, turning her overwhelming workload into something she could manage easily.

3. Reframing Beliefs and Shifting Focus

With these quantum ideas in mind, Neha began to challenge her limiting beliefs about work-life balance. She realized that her belief that balance was impossible was just one possible way of thinking. By shifting her focus and believing that balance was achievable, she started to break through her mental barriers, just like particles tunneling through obstacles.

4. Practical Steps: Prioritization and Setting Boundaries

Using the idea of decoherence, Neha started to collapse the overwhelming possibilities into a clear and manageable reality by taking practical steps. She prioritized her tasks, focusing on what was most important in both her work and personal life. She set clear boundaries, like not checking work emails after a certain time, and made sure to dedicate time to self-care and family. These actions helped her create a more balanced and structured daily routine.

5. Delegation and Communication

Neha also realized that part of creating balance involved asking for help when needed. She talked to her team and managers about her workload and started delegating tasks more effectively. This not only reduced her stress but also empowered her team members, creating a more supportive work environment.

6. Embracing Wellness Practices

Neha added wellness practices like meditation, yoga, and regular exercise to her routine. These practices helped her stay grounded and manage stress better, supporting her journey toward a balanced life. She also made time for activities that brought her joy, reinforcing the stability and balance she was working to achieve.

7. Living with Quantum Awareness

Neha started living with a new awareness, seeing her life as part of a larger, interconnected reality. By aligning her thoughts and actions with the principles of quantum mechanics—overcoming barriers through tunneling and creating coherence through intentional choices—she was able to balance her work and personal life with greater ease and harmony.

Overcoming Limitations and Taking Practical Steps

Neha realized that by embracing the ideas of quantum tunneling and decoherence, she could overcome the limitations she had placed on herself. She saw that achieving work-life balance wasn't about having fewer responsibilities or better circumstances, but about approaching her challenges with a new mindset and taking practical steps to create the life she wanted.

Key Insight:

The concepts of quantum tunneling and decoherence teach us that mental and emotional barriers can be overcome, and a balanced reality can be created through intentional choices. Neha's story shows how applying these quantum principles—challenging limiting beliefs, setting priorities, and making conscious decisions—can lead to a more balanced and fulfilling life. By believing in the possibility of change and taking deliberate steps, anyone can break through the barriers that hold them back and create a life that is harmonious and well-balanced.

CHAPTER 14

Quantum Processes and the Conscious Brain

Introduction

"The brain is wider than the sky."

– Emily Dickinson

Imagine if you could tap into the hidden forces of the universe to turn your dreams into reality. Modern science, with its deep insights, holds the keys to transforming how we pursue our goals. In this chapter, we'll explore the fascinating connections between quantum processes and consciousness, showing how a deeper awareness of your thoughts and actions, combined with an understanding of these scientific ideas, can help you achieve what you desire. We'll dive into concepts like the Butterfly Effect, the Zeno Effect, and Chaos Theory, and learn how to apply these ideas in practical ways through small, purposeful steps. By grasping these powerful principles, you can begin to shape your reality and create the life you've always dreamed of. Let's start by exploring the Butterfly Effect, a concept that shows how even the smallest actions can lead to significant outcomes.

The Science of the Butterfly Effect: How Small Changes Can Make a Big Difference

The Butterfly Effect is an intriguing idea that comes from chaos theory, which studies how complex systems behave in unpredictable ways. The Butterfly Effect suggests that small, seemingly insignificant actions can lead to profound and unexpected outcomes over time. The name comes from the idea that a butterfly flapping its wings in one part of the world could eventually cause a tornado in another, illustrating how minor changes in initial conditions can have far-reaching consequences.

How It All Started

The Butterfly Effect was first noticed by a meteorologist named Edward Lorenz in the early 1960s. Lorenz was conducting weather simulations when he observed something surprising. He found that a tiny change in the starting numbers of his weather model led to vastly different weather predictions. For instance, changing a number from 0.506127 to 0.506—a seemingly minor adjustment—resulted in completely different weather patterns. This discovery highlighted the sensitivity of complex systems to initial conditions and showed why making long-term predictions in such systems is challenging.

Why Small Changes Matter

The importance of the Butterfly Effect lies in its demonstration that small actions can have significant impacts, especially in systems that are complex and unpredictable. These systems, unlike simple ones where small inputs lead to small outputs, can amplify minor variations into major outcomes. This concept is applicable not only in weather forecasting but also in many aspects of life, including human behavior, economics, and even our thoughts.

Examples from Indian History

One of the most striking examples of the Butterfly Effect in Indian history is Mahatma Gandhi's Dandi March in 1930. What began as a small, peaceful protest against the British salt tax eventually sparked a nationwide movement for independence. Gandhi's act of making salt at the seashore might have seemed insignificant at the time, but it led to a series of events that changed the course of Indian history, culminating in India's independence in 1947. This demonstrates how a seemingly small action can ripple through time and bring about monumental changes.

Similarly, the decision by Rani Lakshmibai of Jhansi to resist British rule during the 1857 revolt might have appeared as an isolated act of defiance. However, her courage inspired others and contributed to the broader struggle for Indian independence, making her a symbol of resistance and empowerment.

In the context of the brain and consciousness, the Butterfly Effect suggests that small, intentional changes in your thoughts or behaviors can lead to significant transformations in your life. By making conscious, positive choices, you can set off a chain reaction that leads to greater personal growth and fulfillment.

Quantum Processes: The Foundation of Conscious Thought

At the heart of quantum mechanics is the idea that particles can exist in multiple states at once—a concept known as superposition. These particles remain in this state of possibilities until they are observed or measured, at which point they "collapse" into a single state. This principle is not just a curious fact of the quantum world; it has profound implications for understanding how our thoughts and intentions can influence reality.

Quantum Processes in the Brain

In the context of the brain, quantum processes might be at work in the microtubules of neurons, which could operate at the quantum level to process information in ways that classical physics cannot fully explain. These quantum processes might be responsible for the emergence of consciousness itself, integrating information from various parts of the brain to create a cohesive experience of reality.

Historical Example: Swami Vivekananda's Focused Thought

Swami Vivekananda, one of India's greatest spiritual leaders, exemplified the power of focused thought. His ability to concentrate deeply on spiritual and philosophical ideas led him to insights that inspired millions. Vivekananda's famous speech at the Parliament of the World's Religions in Chicago in 1893, where he introduced Hindu philosophy to the Western world, is an example of how focused thought can influence and shape reality. His clear vision and deep understanding of Vedanta made a significant impact globally, much like how quantum processes in the brain can influence our perception and reality.

The Zeno Effect: Freezing Reality with Focused Attention

The Zeno Effect is a fascinating concept from quantum mechanics that shows how the act of continuous observation can "freeze" a quantum system in its current state, preventing it from evolving. Imagine a situation where you're constantly watching a pot of water, waiting for it to boil. It feels like it takes forever, right? This is a bit like the Zeno Effect: the more you observe something closely, the more it seems to stay the same.

In the quantum world, particles can exist in a superposition of states, meaning they can be in multiple conditions at once. However, when we observe them, we force them to "choose" a state, collapsing all possibilities into one. If we keep observing repeatedly, we can actually stop the system from changing or evolving, effectively "freezing" it in place.

Application in Daily Life: The Zeno Effect can be applied to our lives in the sense that if you focus intensely and consistently on a goal or thought, you can keep it "alive" in a state of potential, making it more likely to happen. For example, if you have a dream or a goal, and you keep it in your mind constantly—thinking about it, planning for it, working towards it—it's like you're not letting the possibility fade away. You're "freezing" it in place, keeping the potential alive until it becomes reality.

Historical Example: Dr. A.P.J. Abdul Kalam's Vision for India

Dr. A.P.J. Abdul Kalam, known as the "Missile Man of India," exemplified the Zeno Effect through his unwavering focus on India's self-reliance in space and defense technologies. Despite numerous challenges, his persistent attention to his vision kept the dream alive, eventually leading to significant advancements in India's space and missile programs. Dr. Kalam's ability to "freeze" his vision in place through continuous focus and effort made it a reality, showing how the Zeno Effect can be applied to achieve long-term goals.

Chaos Theory: The Unpredictability of Complex Systems

Chaos Theory is a fascinating concept that helps us understand how small changes at the beginning of a process can lead to big, often unpredictable results later on. Imagine it like this: even if a system follows certain rules, like how the weather changes or how the stock market works, it's still incredibly difficult to predict what will happen in the long run. This is because tiny differences in the starting conditions can grow and multiply over time, leading to completely different outcomes.

Let's say you're trying to predict the weather for next week. Even if you have all the information about today's weather, like the temperature, humidity, and wind speed, a small difference in any of these details can make a huge difference in what the weather will be like a week from now. For example, a tiny change in wind speed today might lead to a completely different weather pattern next week. This is why weather forecasts are often accurate only for a few days ahead; beyond that, the predictions become less reliable because of how these small changes add up.

This idea can be applied to many areas of life. For instance, think about how a small decision you make today, like choosing to study an extra hour, could lead to significant changes in your future, like getting into a better college or landing a great job. On the other hand, deciding to skip that study session might lead to missing an important opportunity later on. Chaos Theory tells us that even the smallest actions or choices can have a big impact down the road, sometimes in ways we can't predict.

A Deeper Look: Chaos Theory shows us that life is full of complexities and uncertainties. Even when we think we have control over a situation, small factors we might not even notice can change everything. This is true for natural systems, like the weather, as well as human-made systems, like the economy.

The key takeaway from Chaos Theory is that while we can make educated guesses about what might happen, we should always be prepared for surprises. Life doesn't always follow a straight path, and small, seemingly insignificant actions can lead to major, unexpected outcomes. This understanding encourages us to pay attention to the small details and be mindful of our decisions, knowing that they can shape our future in ways we might not immediately see.

By keeping Chaos Theory in mind, we can better navigate the unpredictability of life, understanding that while we can't always control the outcome, we can influence it with the small choices we make every day.

Examples from Indian History

A powerful example of Chaos Theory in Indian history is the Quit India Movement of 1942. Initially, the movement faced severe repression, and many leaders were imprisoned. However, the arrest of leaders like Gandhi and Nehru triggered widespread protests across the country, which the British could not have predicted. This spontaneous and widespread reaction created a situation that spiraled out of control for the British, eventually contributing to India's independence. This illustrates how small, seemingly isolated actions can lead to large, unpredictable outcomes.

Similarly, the Green Revolution in India during the 1960s is an example of how small changes in agricultural practices led to massive changes in food production, economy, and society. The introduction of high-yielding varieties of seeds, combined with irrigation and fertilizers, transformed India from a food-deficient country to one of the world's largest agricultural producers. However, this revolution also brought unforeseen challenges, such as environmental degradation and socioeconomic disparities, demonstrating the unpredictable nature of complex systems.

Consciousness as a Quantum Observer: The Role of Intent and Awareness

In quantum mechanics, the observer effect states that the act of observation can influence the outcome of a quantum event. This idea can be extended to consciousness, where your intent and awareness act as the "observer" that shapes your reality. By consciously directing your thoughts and actions, you can influence the potential outcomes in your life.

Application in Daily Life: Your focused thoughts, intentions, and awareness can act as the "observer" of your reality. By directing these consciously, you might influence which possibilities manifest, much like how observing a quantum system determines its state. This suggests that you have the power to shape your reality through conscious thought and deliberate action.

Historical Example: Sardar Vallabhbhai Patel's Vision for a United India

Sardar Vallabhbhai Patel, known as the "Iron Man of India," played a crucial role in uniting over 500 princely states into the newly independent Indian Union. Patel's focused intent on creating a united India acted as a quantum observer, shaping the nation's reality. His clear vision and determined actions prevented the fragmentation of India and ensured its integrity as a sovereign state. Patel's efforts demonstrate how focused intent and awareness can shape reality and bring about profound national change.

Quantum Resonance: Tuning Your Mind to the Universe

Quantum Resonance is a concept that explains how different quantum systems can "resonate" or synchronize with each other, amplifying their effects. Imagine two tuning forks. If you strike one, the other, if it's tuned to the same frequency, will start vibrating too, even if you didn't touch it. This is resonance—when things vibrate in harmony, they amplify each other.

In the context of the brain, Quantum Resonance suggests that when our thoughts and emotions are aligned and in harmony, they resonate with each other, creating a powerful mental state. This resonance can help integrate various aspects of our consciousness—like our memories, sensory experiences, and emotions—into a unified whole, leading to a clearer and more focused mind.

Application in Daily Life: When your thoughts, emotions, and actions are in alignment with your goals, they resonate together, amplifying your ability to achieve what you desire. For instance, if you're working towards a goal, and you think positively about it, feel motivated, and take actions that align with it, you create a powerful force that drives you towards success. It's like getting all parts of your mind to "sing" in harmony, making your intentions stronger and more likely to manifest.

Historical Example: Rabindranath Tagore's Resonance with Indian Culture

Rabindranath Tagore, India's celebrated poet and philosopher, exemplified Quantum Resonance through his deep connection with Indian culture and spirituality. His thoughts, emotions, and creative expressions were in perfect harmony with the cultural and spiritual heritage of India. This resonance amplified his influence, making his work deeply impactful not only in India but across the world. Tagore's alignment with the values and spirit of India created a powerful resonance that continues to inspire generations, demonstrating how tuning your mind to a higher purpose can amplify your impact.

The Quantum Brain: A New Frontier in Consciousness Studies

Recent theories propose that the brain's microtubules might function as quantum computers, processing information in ways that classical physics cannot fully explain. These quantum processes could be responsible for the emergence of consciousness, integrating information from various parts of the brain to create a cohesive experience of self and reality.

Application in Daily Life: If the brain operates on quantum principles, our thoughts and intentions might be deeply connected to the quantum nature of the universe. This means that our mental processes are not just the result of neural activity but are also influenced by quantum processes that connect us to the broader universe. By understanding and leveraging these quantum processes, we can enhance our cognitive abilities and better align our consciousness with the universal super-consciousness.

Historical Example: Aryabhata's Pioneering Work in Mathematics and Astronomy

Aryabhata, one of India's greatest mathematicians and astronomers, laid the foundation for many modern scientific principles. His deep understanding of mathematics and his pioneering work in astronomy demonstrate how a mind attuned to the universe can uncover fundamental truths. Aryabhata's work, much like the exploration of quantum processes in the brain, opened new frontiers in our understanding of the world and the cosmos. His contributions continue to resonate in the fields of science and mathematics, showing how a well-aligned mind can achieve extraordinary insights and advancements.

Conclusion

In this chapter, we explored how quantum processes and consciousness are intricately linked, offering powerful insights into how we can influence our reality. By understanding concepts like the Butterfly Effect, the Zeno Effect, and quantum resonance, we can harness the power of our minds to shape our lives and achieve our goals. The quantum brain represents a new frontier in consciousness studies, suggesting that our thoughts and intentions may be deeply connected to the fabric of the universe. As you continue your journey of exploration, remember that even the smallest thoughts and actions can have profound effects on your reality, just as quantum processes shape the universe at its most fundamental level.

Through examples from Indian history, we see how these principles have been applied to bring about significant change. Whether it's through the focused intent of leaders like Gandhi and Patel or the resonance of teachings from Vivekananda and Aryabhata, the lessons of quantum processes and consciousness offer a powerful guide for shaping your future and connecting with the greater universe.

Story Time: Rahul's Struggle with Lust and Cheating

Rahul was a young and successful executive. He had a great job, people admired him, and it seemed like he had everything going well. But deep down, Rahul had a problem—he couldn't control his desires, especially when it came to other women. Even though he was in a serious relationship with his girlfriend, Rahul often felt attracted to other women and sometimes even cheated on her. He knew this was wrong and felt terrible every time, but he couldn't stop himself. Rahul believed his desires were too strong and that he was stuck in a cycle of cheating and feeling guilty.

The Problem: Lust and Cheating

Rahul's inability to control his desires led him to cheat, which made him feel guilty. His relationship with his girlfriend was suffering, and he felt trapped, thinking he couldn't change.

A New Way of Thinking: The Brain and Consciousness

Rahul wanted to change, so he decided to seek help from a therapist. In therapy, he learned something important: our thoughts and desires are influenced by how our brain works, and sometimes, these thoughts are very complicated. But just because these thoughts come up doesn't mean we have to act on them.

Solution: How Rahul Overcame His Struggles

1. Becoming Mindful of His Thoughts

Rahul started practicing mindfulness, which means paying attention to what's happening in your mind without judging it. Instead of trying to push his desires away, he learned to notice them and understand where they came from. This helped him realize that even though the thoughts came automatically, he didn't have to act on them.

Example: Rahul thought of his mind like a radio. Just because a song comes on doesn't mean he has to listen to it. He could choose which "station" to focus on and ignore the others.

2. Understanding What Triggers His Desires

Through therapy, Rahul discovered that his desire to cheat often came from feelings of stress, loneliness, or insecurity. These emotions would trigger his urges, leading him to make poor decisions. By addressing these feelings directly, Rahul found that his urges became weaker.

Example: Rahul compared this to fixing the foundation of a house. Instead of just patching up cracks, he worked on the underlying issues, so the cracks wouldn't keep appearing.

3. Changing How He Thinks About His Desires

Rahul learned a technique called cognitive behavioral therapy (CBT), which helped him change the way he thought about his desires. Instead of thinking, "I can't resist this," he started telling himself, "I can choose to do the right thing." This made him feel more in control.

Example: It was like learning to drive. At first, it felt overwhelming, but with practice, Rahul realized he could steer his thoughts in the direction he wanted.

4. Strengthening His Relationship

Rahul knew that his relationship with his girlfriend needed work. They started going to couples therapy, where they learned how to communicate better. By being honest about his struggles, Rahul and his girlfriend were able to rebuild trust and create a stronger bond.

Example: Rahul saw this like rebuilding a bridge. Each honest conversation was like adding a new plank, making the bridge stronger so they could connect better.

5. Avoiding Temptation and Building Good Habits

Rahul made a conscious effort to avoid situations where he might be tempted to cheat. He also started creating positive habits, like spending more quality time with his girlfriend and focusing on activities that made him feel good about himself.

Example: Rahul thought of this like planting a garden. By removing weeds (temptations) and nurturing the flowers (good habits), he created a healthier environment for his relationship to grow.

6. Believing He Could Change

Finally, Rahul adopted a growth mindset, meaning he believed that he could change and improve with effort. He understood that overcoming his urges would take time, but he was committed to working on it every day.

Example: Rahul saw this like training for a marathon. Each day of practice brought him closer to his goal, and he knew that with perseverance, he could succeed.

Over time, Rahul noticed that his urges to cheat became less frequent. He felt more in control of his actions, and his relationship with his girlfriend improved. Rahul realized that while his thoughts might be influenced by complicated processes in his brain, he had the power to make better choices. He wasn't trapped by his desires—he could change, and he did.

Key Insight:

Rahul's story teaches us that even though our thoughts and desires can be complicated, they don't control us. By being mindful, understanding what triggers our behavior, and choosing to act differently, we can overcome challenges like Rahul did. It's about taking small steps every day to build better habits and stronger relationships, showing us that change is possible with the right mindset and effort.

CHAPTER 15

The Synchrony of Mind and Cosmos

Introduction

"The cosmos is within us.
We are made of star-stuff. We are a way for the universe to know itself."

– Carl Sagan

Scientists have found that the connection between our minds and the universe might be deeper than we ever thought. Quantum physics shows us that particles are connected in ways that don't make sense according to regular physics. This connection, called quantum entanglement, suggests that everything in the universe, including our thoughts, might be linked together.

Neuroscientists also believe that our brains aren't separate from the world around us. They can be influenced by things like the Earth's electromagnetic fields and other forces from the universe. Just like our brains follow a daily rhythm, they might also be in sync with the universe on a deeper level.

Cosmologists (scientists who study the universe) have discovered that the elements making up our bodies were created in stars. Carl Sagan's quote, "We are made of star-stuff," isn't just a poetic phrase—it's a scientific fact that highlights how we are part of the universe.

By understanding these scientific ideas, we can start to see how our minds and the universe might be connected. Imagine that your mind and the universe are like two dancers moving in perfect harmony. This chapter will explore how the mind and the universe might influence each other, creating a connection that is both deep and meaningful.

The Universe Within: Inner Cosmology

Microcosm and Macrocosm

The saying "As above, so below" means that the patterns we see in the vast universe (the macrocosm) might also be present within us (the microcosm). For example, just as planets orbit the sun, electrons orbit the nucleus of atoms in our bodies. This suggests that we are like small versions of the universe, carrying the same patterns and principles within us.

Example: Think of a snowflake. No two snowflakes are exactly the same, but they all have similar patterns. Similarly, while each person is unique, we all share common structures—like how our brains work—that reflect the larger patterns found in the universe.

Cosmic Resonance and Mind

Harmonic Resonance: Everything in the universe vibrates at certain frequencies, from the smallest particles to the largest planets. The idea of harmonic resonance suggests that our minds might "tune in" to these cosmic vibrations, creating moments when our thoughts and the universe seem perfectly aligned. This might explain why we sometimes experience coincidences that feel too meaningful to be random—these could be times when our minds are resonating with the universe.

Example: Imagine listening to a radio. When you tune it to the right frequency, you clearly hear your favorite song. In the same way, your mind might "tune" into the universe, catching signals that help guide your thoughts and actions in ways that feel almost magical.

Celestial Rhythms: The universe works in cycles—day and night, the phases of the moon, and the changing seasons. These cycles can influence how we feel and act. For example, many people feel more emotional or energized during a full moon. By paying attention to these cycles, we might better understand our own moods and actions.

Example: Think about how you feel during different times of the year. Many people feel more energetic in the spring and summer, while the shorter days of winter might bring on feelings of tiredness or sadness. These changes in mood could be linked to the natural cycles of the Earth and the sky.

Mind as a Creative Force in the Cosmos

1. Cosmic Creativity and the Mind

Universal Imagination: Creativity isn't just something humans do—it's a force that drives the entire universe. From the formation of stars to the creation of art, creation is a universal principle. When you imagine a new idea, you are tapping into the same creative force that shaped the cosmos. This means that by creating, we are participating in the universe's ongoing story of creation.

Example: Think of a painter creating a masterpiece. The painter starts with a blank canvas, just like the universe started with nothing before the Big Bang. Every brushstroke adds to the painting, just as every event in the universe adds to its story. Your thoughts and creations are like brushstrokes on the canvas of the universe.

Manifestation of Ideas: Bringing ideas from your mind into reality—like writing a book, building a business, or solving a problem—is similar to how the universe brings matter into existence. This means that your mind has the power to shape reality, not just in your personal life, but as part of the greater cosmic process.

Example: Imagine you have an idea for a new invention. At first, it only exists in your mind. But as you start planning and building, that idea begins to take shape in the real world. This process mirrors how the universe takes energy and turns it into matter—turning potential into reality.

2. Mind and the Evolution of the Universe

Conscious Evolution: Just as the universe is constantly evolving, becoming more complex over time, so too is human consciousness. This suggests that our thoughts and awareness are part of the universe's evolution. As we learn and grow, we contribute to the ongoing development of the cosmos.

Example: Consider how human society has evolved over time—from simple tribes to complex civilizations. Our growing knowledge and understanding of the world are part of the universe's story of becoming more complex and aware.

Creative Evolution: The idea that the universe is not just evolving physically, but also in terms of consciousness and creativity, suggests that your mind plays a role in the universe's growth. Every time you think creatively or solve a problem, you're participating in the universe's journey toward greater complexity and understanding.

Example: Think about the progress of technology over the past century. Innovations like the internet, smartphones, and space travel didn't just happen—they were imagined and created by people like you, using their minds to push the boundaries of what's possible. This creativity is a continuation of the universe's creative energy.

The Interplay of Thought and Cosmic Energy

1. Cosmic Energy Fields and Human Thought

Interaction with Cosmic Energy: The universe is filled with invisible energy fields that affect everything from how planets move to how particles interact. Your mind might also interact with these cosmic energy fields. Thoughts could be seen as a form of energy, and just like a pebble creating ripples in a pond, your thoughts might send ripples through the universe's energy fields.

Example: Have you ever noticed how your mood can affect the atmosphere of a room? When you're happy and positive, others around you might start to feel uplifted as well. This could be an example of your mental energy influencing the environment, much like how cosmic energy fields influence matter in the universe.

Chi and Prana in a Cosmic Context: Concepts like Chi (in Chinese philosophy) or Prana (in Indian philosophy) refer to vital life forces that flow through all living beings. When we think of these forces in a cosmic context, they become part of the universe's larger energy fields, connecting our individual consciousness with the greater whole.

Example: Imagine you're practicing yoga and focusing on your breath. As you breathe in and out, you're not just taking in oxygen; you're also drawing in energy from the universe. This energy helps you feel balanced and connected, showing how our personal well-being is linked to the energy of the cosmos.

2. Mind-Driven Cosmic Phenomena

Cosmic Co-Creation: Some thinkers suggest that consciousness might play a role in the formation or behavior of cosmic phenomena. This idea, though speculative, proposes that the collective consciousness of humans—or even all living beings—could influence how the universe develops, like shaping new planets or life forms.

Example: Imagine a large group of people coming together with a shared intention, like praying for peace. While it's hard to measure, some believe that such collective consciousness could send out positive energy that might influence events in the world or even in the cosmos, much like how gravity pulls objects together.

Thought Forms as Cosmic Entities: In certain spiritual traditions, thoughts are believed to have a life of their own, influencing the material world. When expanded to a cosmic scale, it's possible to imagine that our collective thoughts could help shape the universe itself, contributing to the creation of new realities.

Example: Consider the power of a shared dream, like the dream of exploring space. This idea began as a thought in the minds of a few visionaries, but over time, it grew into a reality, with humans landing on the moon and sending spacecraft to other planets. Our thoughts can indeed have far-reaching effects, shaping not just our world, but potentially the cosmos.

The Cosmos as a Conscious Entity

1. Cosmic Consciousness

The Universe's Mind: Some believe that the universe itself is a living, conscious entity, with each individual mind being a tiny part of this vast consciousness. This idea suggests that when you think, feel, or dream, you are participating in the thoughts and dreams of the universe.

Example: Imagine the universe as a giant brain, with each star, planet, and person acting like a neuron in that brain. Your thoughts and experiences are like the signals passing between these neurons, contributing to the universe's overall awareness.

Gaia Hypothesis Expanded: The Gaia Hypothesis suggests that Earth functions as a single living organism. Expanding this idea to the entire universe means considering the cosmos as a living, thinking entity that evolves and grows, just like a giant cosmic body.

Example: Think about how your body functions. Each cell has a role, working together to keep you alive and healthy. Now, imagine that the universe works the same way, with planets, stars, and even human minds acting as the "cells" that keep the cosmic body alive and evolving.

2. Universal Intuition

Tapping into Cosmic Wisdom: Intuition, that gut feeling or inner knowing, might be more than just a subconscious processing of information. Some believe it's a way of tapping into the universal consciousness—a direct line to the wisdom of the cosmos. When you have a sudden insight or a strong feeling about something without knowing why, it could be because your mind is briefly aligning with the universe's vast store of knowledge.

Example: Have you ever had a hunch that turned out to be correct, even though there was no logical reason for you to know the answer? This could be an example of universal intuition, where your mind briefly connects with the cosmic consciousness to access knowledge that's normally beyond our reach.

Dreams as Windows to the Cosmos: Dreams are often seen as personal experiences, but what if they're also a way for our minds to connect with the universe? Some theories suggest that during sleep, when our conscious mind is at rest, we may be more open to receiving messages or insights from the cosmos, allowing us to explore possibilities beyond our waking reality.

Example: Consider a dream where you're flying through space or meeting beings from another world. While it might seem like pure fantasy, some believe these dreams could be your mind's way of exploring the cosmos, connecting with different aspects of universal consciousness.

Practical Applications: Living in Sync with the Universe

1. Mindful Alignment with Cosmic Cycles

Living by Natural Rhythms: By aligning our daily lives with the natural rhythms of the universe—such as the lunar phases, seasons, and solar cycles—we can live more harmoniously with the world around us. This alignment can bring a sense of peace and balance, helping us feel more connected to the larger cosmic order.

Example: Consider planting a garden according to the lunar cycle. Many gardeners believe that planting seeds during a new moon and harvesting during a full moon aligns with natural growth rhythms, resulting in healthier plants. Similarly, organizing your life around these natural cycles might help you feel more in tune with the universe.

2. Cosmic Meditation and Visualization

Meditating on the Universe: Meditation that focuses on the cosmos—such as visualizing yourself as part of a vast, interconnected universe—can help you feel more connected to the greater whole. This practice can expand your awareness and help you tap into the calming, expansive energy of the cosmos.

Example: During meditation, imagine yourself floating in space, surrounded by stars. Feel the vastness of the universe and recognize that you are a part of this infinite expanse. This can help you put daily worries into perspective, reducing stress and enhancing your sense of peace.

Creative Visualization for Cosmic Harmony: Use creative visualization techniques to imagine positive changes not just in your life but in the world around you. By focusing on harmonious outcomes, you might contribute to creating a more peaceful, balanced environment.

Example: If you're facing a difficult situation, try visualizing a peaceful resolution that benefits everyone involved. Imagine the energy of this resolution spreading out, like ripples in a pond, influencing not just your life but the lives of others in a positive way.

Conclusion

The idea that our minds are connected to the universe in profound ways opens up exciting possibilities for understanding our place in the cosmos. By exploring how our thoughts, emotions, and actions interact with the larger cosmic forces, we can begin to see ourselves as active participants in the universe's grand dance. This connection not only enriches our understanding of who we are but also empowers us to live more intentionally, aligning our lives with the rhythms and energies of the cosmos.

As you reflect on these concepts, consider how you might apply them in your daily life. Whether through meditation, mindful living, or simply being more aware of your connection to the universe, you have the opportunity to tap into the boundless potential of the cosmos, shaping your reality in harmony with the greater whole.

Story Time: Samir's Journey to Health and Balance

Samir was a young professional with a bright future. He loved staying active, playing sports, and enjoying the outdoors. But in his late twenties, Samir was diagnosed with a chronic autoimmune disease. This illness made him feel constantly tired and in pain. The doctors told him that while medicine could help manage his symptoms, there was no cure, and he would likely have to live with this condition for the rest of his life.

This news was very hard for Samir. He had to give up many activities he loved, and his job began to suffer because he often needed time off. Samir felt like his body had betrayed him, leading him into deep sadness and depression. He saw his body as the enemy, holding him back from living the life he had dreamed of. He felt angry, frustrated, and hopeless, believing that his illness was a barrier to his happiness and success.

The Problem: Health Issues and the Mental Struggle with Chronic Illness

Samir's chronic illness not only affected his physical abilities but also caused him a lot of emotional pain. He felt trapped in a body that no longer supported his goals, which deeply impacted his mental health and overall quality of life.

The Solution: Combining Medical Treatment with Mindfulness and Positive Thinking

As time went on, Samir realized that besides taking his prescribed medicine, he needed to find ways to take care of his mental and emotional health. He started looking into other practices that could help him feel more in control of his life, focusing on the power of his mind along with his medical care.

Understanding the Connection Between Mind and Body: Samir learned that the mind and body are closely connected. How you think and feel can affect your physical health. He realized that while his medical treatment was essential, managing his stress and staying positive could also help improve his health. This gave him hope and motivated him to take active steps to support his healing.

Adopting a Positive Mindset: Determined to take control of his life, Samir began practicing positive thinking and visualization techniques. Every day, he imagined his body responding well to the treatment and getting stronger. He also started telling himself, "I am strong, and I can handle my health challenges." These practices helped him focus on what he could still do, without forgetting the importance of his medical treatment.

Incorporating Mindfulness and Meditation: Samir added mindfulness and meditation to his daily routine. These practices helped him manage stress and improve his mental well-being. Mindfulness allowed him to stay present, reducing his worries about the future and helping him cope with the daily challenges of his condition. Through meditation, Samir learned to observe his thoughts and emotions without getting overwhelmed, which helped him stay calm and balanced.

Balancing Medical Help with Mindfulness Practices: Samir continued following his doctor's advice, taking his medications regularly, and attending all his medical appointments. He understood that mindfulness, positive thinking, and meditation worked best when combined with his medical care. By using both approaches, Samir found a balance that helped him manage his condition and improve his quality of life.

Living with Purpose: As Samir's mindset changed, he stopped seeing his illness as an enemy. Instead, he viewed it as a challenge he could manage. He found new ways to live a fulfilling life within his physical limitations, focusing on what he could do rather than what he couldn't. Samir also began sharing his journey with others facing similar challenges, offering hope and advice on how to use mindfulness and positive thinking alongside medical treatment. This gave him a renewed sense of purpose, which further improved his mental and physical health.

Key Insight:

Samir's story shows the importance of combining medical treatment with mindfulness and positive thinking. While the mind can help manage stress and improve mental health, it should work alongside, not replace, conventional medical care. By using both the science of medicine and the benefits of mental practices, people like Samir can find a balanced, fulfilling way to live with chronic illness. This approach empowers individuals to take control of their health and well-being, helping them become stronger and more resilient in the face of challenges.

CHAPTER 16

The Holonomic Brain, Morphic Resonance, and Observing Consciousness

Introduction

*"The universe is not outside of you. Look inside yourself;
everything that you want, you already are."*

– Rumi

The human mind is incredibly complex, and even though we've learned a lot about how the brain works, there's still so much we don't understand. This chapter explores some interesting and lesser-known ideas about how the brain, memory, and consciousness might connect with the universe.

1. The Holonomic Brain: A New Way of Thinking About Memory and Perception

Traditionally, we think of the brain as a kind of computer, with different parts doing different jobs. But the holonomic brain theory, proposed by neuroscientist Karl Pribram and physicist David Bohm, suggests something more interesting. According to this theory, the brain works like a hologram, where each part of the brain holds information about the whole.

Holographic Memory Storage Imagine you tear a photograph in half. Normally, you'd only get half the image. But with a hologram, each piece still shows the entire image, just from a different angle. Pribram suggested that memories in the brain are stored in a similar way. This means that every part of the brain has all the information needed to recall a memory, which might explain why people often retain memories even after brain damage.

Example: Think about how a familiar smell can trigger a whole memory from your childhood. This could be because your brain stores memories like a hologram, with each part holding the full picture.

Perception and the Holographic Model The holographic model also suggests that perception works in a similar way. Instead of processing sensory information one step at a time, the brain might create a holographic image of the world, allowing us to see reality as a whole.

Example: When you look at a painting, you see the entire image, not just individual brush strokes. The holonomic brain theory suggests that this ability to see the whole picture might come from the brain's holographic nature.

2. Morphic Resonance: The Idea of Collective Memory

Biologist Rupert Sheldrake proposed the theory of morphic resonance, which suggests that there's a kind of collective memory in nature, shared by all members of a species. According to this idea, once something is learned by one individual, it becomes easier for others to learn it too, as if the knowledge is being shared across an invisible network.

Collective Memory and Learning Morphic resonance suggests that when one person learns something new, this information becomes available to others through a shared field of memory. This could explain why scientific discoveries or new ideas often appear independently in different parts of the world at the same time.

Example: Think about how different cultures developed similar technologies, like the wheel, even though they were far apart. Sheldrake's theory suggests that once one person learns something, it becomes easier for others to learn it too, as if knowledge is being shared through a collective memory.

Habitual Patterns in Nature Morphic resonance might also explain why certain behaviors or patterns in nature repeat across generations, not because they're in the DNA, but because they're passed down through these morphic fields.

Example: Think of migratory birds that follow the same routes every year, even if they've never traveled them before. According to morphic resonance, these birds might be tapping into a collective memory that guides them.

3. Observing Consciousness: How the Mind Might Shape Reality

In this final section, we explore how consciousness might not just observe reality but actually shape it. While quantum mechanics often comes up in discussions of consciousness, there's also a more practical way to think about how our minds interact with the world.

The Role of Attention Attention is powerful. When we focus on something, we're not just watching it—we're engaging with it in a way that can change our experience and even the outcome. This idea suggests that consciousness, through focused attention, might play a key role in shaping reality.

Example: When you focus on learning a new skill, like playing an instrument, your brain starts to change to improve your ability. This isn't just about practice—it's about how your focused attention actively changes your brain and possibly your reality.

Consciousness and Reality Formation There's growing evidence that our beliefs, expectations, and mental focus can influence our physical reality. This could be seen as a way that consciousness shapes reality in everyday life.

Example: Think about the placebo effect, where believing in a treatment can lead to real health improvements, even if the treatment has no active ingredients. This shows how consciousness can directly influence physical outcomes.

The Power of Collective Consciousness If individual consciousness can shape reality, what about collective consciousness? The idea here is that when large groups of people focus on the same goal or idea, they might be able to bring about real change in the world.

Example: Mass meditation events, where many people meditate on peace or healing, have been reported to lower crime rates and violence. While these claims are debated, they hint at the potential power of collective consciousness.

Application in Daily Life

1. Overcoming Laziness and Procrastination

Holonomic Brain: Visualizing Success The holonomic brain theory suggests that focusing on one positive aspect of your goal can help activate the whole picture of success.

Step: Create a clear mental image of achieving your goal. Visualize yourself completing a task with enthusiasm. This engages the brain's memory system, making success feel more achievable.

Application: When you feel procrastination coming on, close your eyes and visualize completing your task. This can make the task feel more manageable and less overwhelming.

Morphic Resonance: Tapping into Collective Willpower Surround yourself with disciplined, productive people, either in person or online. Their success can influence your behavior.

Step: Join a study group or an online community that values hard work. This can help you tap into their collective discipline.

Application: Participate in a productivity challenge or work with others who share your goals. Their energy can help you overcome laziness.

Observing Consciousness: Focusing Attention Direct your attention to the task at hand to reduce procrastination.

Step: Break your work into smaller tasks and focus intently on each one. This helps you stay engaged.

Application: Use the Pomodoro Technique—work for 25 minutes, then take a break. This focused effort can help you push through procrastination.

2. Maintaining Discipline

Holonomic Brain: Repeating Positive Routines By repeating disciplined actions daily, you create strong memory patterns that make it easier to maintain discipline.

Step: Start with small, manageable tasks that you can repeat every day.

Application: If you're trying to exercise every morning, start with just 5 minutes. Each time you do it, you reinforce the habit.

Morphic Resonance: Resonating with Disciplinary Fields Spend time with people who have strong self-discipline. Their habits can influence you.

Step: Identify role models or communities that value discipline and interact with them regularly.

Application: Join a fitness group or follow productivity experts. Their discipline can help you maintain your own.

Observing Consciousness: Mindful Discipline Focus on the benefits of discipline to strengthen your commitment.

Step: Remind yourself regularly of the long-term benefits of discipline.

Application: Keep a journal to track your disciplined actions and reflect on your progress.

3. Starting a New Habit

Holonomic Brain: Creating a Strong Initial Memory Start your new habit with a positive experience to make it easier to repeat.

Step: Make the first few times you do the new habit enjoyable and rewarding.

Application: If you're starting a meditation practice, choose a peaceful spot and add calming music. This will help you form a positive memory of the habit.

Morphic Resonance: Aligning with Habitual Fields Align yourself with people who already practice the habit you want to start.

Step: Join a community where your desired habit is common.

Application: If you want to start reading daily, join a book club or an online reading challenge.

Observing Consciousness: Conscious Commitment Set your intention clearly and focus on the new habit.

Step: Dedicate time each day to your new habit and focus your attention on it.

Application: Before starting a new exercise routine, take a moment to mentally commit to it. This will help anchor the habit in your daily life.

4. Breaking Old Habits

Holonomic Brain: Replacing Negative Patterns Replace an old habit with a new, positive one.

Step: Identify the old habit and consciously choose a new behavior to replace it.

Application: If you're trying to quit smoking, replace it with chewing gum or taking a walk.

Morphic Resonance: Disconnecting from Negative Fields Distance yourself from environments and people that reinforce the old habit.

Step: Change your environment and spend time with people who model the behavior you want to adopt.

Application: Avoid places and social circles that encourage the habit you're trying to break.

Observing Consciousness: Redirecting Focus Shift your focus away from the old habit and towards the benefits of breaking it.

Step: Each time you feel the urge to engage in the old habit, redirect your focus to the positive outcomes of breaking it.

Application: If you're trying to reduce social media use, focus on a productive task instead.

5. Getting Out of Guilt

Holonomic Brain: Reframing Past Experiences Reframe your past experiences in a more positive light to reduce guilt.

Step: Focus on what you've learned from the experience and how it has helped you grow.

Application: If you feel guilty about a past mistake, think about how it has made you a better person today.

Morphic Resonance: Connecting with Forgiveness Surround yourself with people who have moved on from guilt and resonate with their emotional state.

Step: Engage with communities or support groups that emphasize forgiveness and self-compassion.

Application: Join a support group and learn from others who have overcome guilt.

Observing Consciousness: Mindful Self-Compassion Observe your guilt without judgment and focus on self-compassion.

Step: Practice mindfulness by acknowledging your guilt without attaching to it.

Application: When guilt arises, take a moment to sit quietly and observe it without reacting.

6. Overcoming Anger Issues

Holonomic Brain: Rewiring Anger Triggers Change how you react to anger triggers to rewire your brain.

Step: Identify your triggers and consciously work to change your response.

Application: If being stuck in traffic makes you angry, practice staying calm and focusing on something positive instead.

Morphic Resonance: Aligning with Calmness Spend time with calm, collected people to help you adopt similar behavior.

Step: Engage with individuals who manage anger well and learn from them.

Application: Watch videos or read books by experts in anger management to help you internalize a more peaceful approach.

Observing Consciousness: Practicing Detachment Observe your anger without immediately acting on it.

Step: When you feel anger rising, pause and observe the emotion without judgment.

Application: When you feel anger building, take a deep breath and count to ten before responding.

Conclusion

In this chapter, we looked at three interesting ideas that give us new ways to think about the brain, memory, and consciousness. We explored the idea that the brain might work like a hologram, where each part holds information about the whole. We also considered the concept of a collective memory, shared by all living beings, through something called morphic resonance. Lastly, we discussed how our consciousness might have the power to shape our reality.

These ideas may seem unusual or even hard to believe, but they encourage us to think beyond the usual scientific explanations. They suggest that our mind and the universe are more connected than we might have thought. As science continues to study the mysteries of the brain and consciousness, these theories might help us better understand the true nature of reality.

Story Time: Anil's Journey from Self-Centeredness to Empathy

Introduction

Anil was a very ambitious corporate executive, known for his sharp mind and determination. His quick rise in the corporate world earned him respect, but it also had a cost. Anil's focus on success made his colleagues see him as arrogant and self-centered. His personal relationships suffered too, as he often put his career above everything else. Anil believed that being self-focused was necessary for success, but deep down, he felt lonely and unfulfilled.

Problem: Ego and Lack of Empathy

Anil's strong ego and self-centered approach created big problems in his relationships, both at work and at home. His lack of empathy led to conflicts and made him feel isolated, even though he was successful on the outside.

Solution: Embracing New Ways of Thinking

As Anil's feelings of emptiness grew, he decided to explore new ways to improve his relationships and find more meaning in life. He attended a workshop on consciousness and personal growth, where he learned about three important ideas: the Holonomic Brain, Morphic Resonance, and Observing Consciousness. These ideas helped Anil understand how he could move past his ego-driven behavior and develop real connections with others.

1. Understanding the Holonomic Brain: Seeing the Bigger Picture

Anil learned that the Holonomic Brain theory suggests that the brain works like a hologram, where each part contains information about the whole. This means our brains can think in a more connected and holistic way. Anil realized that he could develop empathy by seeing things from different perspectives.

Application: Anil started practicing mental exercises where he imagined different scenarios from other people's viewpoints. For example, before making a decision at work, he would think about how it might affect his colleagues. This helped him make decisions that were more considerate of others' feelings and needs.

2. Harnessing Morphic Resonance: Learning from Others

The idea of Morphic Resonance suggests that there is a shared memory or field that connects all living beings. Anil realized that by connecting with this collective memory, he could learn from others and better understand human emotions and relationships.

Application: Anil began surrounding himself with positive influences—books, mentors, and communities that valued empathy and teamwork. By doing this, he naturally started to absorb these values, which helped him build stronger, more empathetic relationships.

3. Observing Consciousness: Becoming Self-Aware

Observing Consciousness taught Anil the importance of being self-aware. He learned that by observing his thoughts and emotions without judging them, he could understand his automatic behaviors, especially those driven by ego.

Application: Anil started a daily mindfulness practice, where he spent time each morning reflecting on his thoughts and emotions. He noticed that many of his actions were driven by a need for validation. By becoming aware of this, he could choose to act differently. For instance, instead of reacting defensively to feedback, he started to listen openly, seeing feedback as an opportunity to grow.

4. Practicing Empathy: Visualization and Listening

To further develop empathy, Anil used visualization techniques and practiced active listening. By imagining himself in others' shoes, he could better understand their feelings and perspectives.

Application: When facing a difficult situation at work, Anil would take a moment to think about how his colleagues might feel. This helped him approach situations with more empathy. He also practiced active listening by giving his full attention to others, which improved his relationships.

5. Balancing Assertiveness with Empathy: Finding Harmony

Anil realized that being empathetic didn't mean he had to give up his own needs or opinions. Instead, it meant finding a balance where he could share his ideas while also respecting others' perspectives.

Application: Anil started practicing this balance in meetings by clearly stating his views but also inviting others to share theirs. This approach led to better outcomes and stronger relationships with his team.

6. Transforming Relationships: Creating Positive Change

As Anil continued to apply these ideas, he noticed big changes in his relationships. His colleagues started seeing him as a supportive leader, and his family life improved as he became more emotionally connected.

Application: Anil's empathy allowed him to resolve conflicts more effectively and create a more collaborative work environment. At home, his family noticed that he was more present and attentive, leading to stronger bonds.

7. Shaping Reality through Consciousness

Through the Holonomic Brain theory, Anil understood that his consciousness could influence his reality. By focusing on positive outcomes and keeping a clear mindset, Anil realized he could shape his experiences and environment.

Application: Anil began using visualization techniques daily, focusing on the outcomes he wanted in both his personal and professional life. Over time, he noticed that his reality started to match these visualizations, reinforcing his belief in the power of consciousness.

Key Insight

Anil's story shows how powerful empathy, mindfulness, and self-awareness can be. By embracing these practices, Anil was able to move past his ego and build stronger, more meaningful connections. His journey reminds us that true fulfillment comes not from focusing only on ourselves but from understanding and connecting with the world around us. By applying these lessons in our own lives, we can create more harmonious and fulfilling relationships.

CHAPTER 17

Learning Math to Connect with God

Introduction

"Mathematics is the language in which God has written the universe."

– Galileo Galilei

Mathematics is often seen as a tough and abstract subject, mostly for scholars or academics. But math is more than just numbers and formulas—it is the fundamental language of the universe, a bridge that connects us to the divine order that governs everything. In this chapter, we will explore how learning mathematics can enrich our spiritual lives and help us understand the deep connections between the universe, our existence, and God. By mastering mathematical principles, we not only boost our intellectual skills but also strengthen our connection with the divine essence of the universe.

Maths is Everywhere

Mathematics is not just a subject studied in classrooms; it is the foundation of everything in our universe. From the way planets orbit the sun to the way leaves grow on trees, math governs all natural processes. Whether we realize it or not, we are constantly interacting with mathematics in our daily lives. It's in the rhythm of our heartbeat, the symmetry of a snowflake, and the patterns in a seashell. By recognizing the presence of math in these everyday occurrences, we can begin to see how deeply connected we are to the universe and, ultimately, to God.

Fibonacci Sequence

The Fibonacci sequence is a series of numbers where each number is the sum of the two preceding ones, starting from 0 and 1. It looks like this: 0, 1, 1, 2, 3, 5, 8, 13, 21, and so on. (0+1=1, 1+1=2, 1+2=3, 2+3=5, 3+5=8, 5+8=13…) This sequence is found in various natural phenomena, demonstrating the inherent mathematical order in the universe.

1. Flower Petals:

Petal Count: Many flowers have a number of petals that is a Fibonacci number. For example:

- Lilies have 3 petals.

- Buttercups have 5 petals.

- Daisies can have 34, 55, or even 89 petals. These numbers are all part of the Fibonacci sequence.

Spiral Patterns: The arrangement of petals around the center of the flower often follows a spiral pattern. The reason behind the Fibonacci sequence in flower petals is related to how plants grow. By following this pattern, plants ensure that each leaf or petal is positioned in a way that minimizes overlap with others, maximizing exposure to sunlight and rain. This is especially important for the flower's reproductive success. The spirals can be counted in one direction and the other, and these counts often match Fibonacci numbers.

Examples in Nature:

- **Sunflowers:** The number of spirals in the seeds at the center of a sunflower often matches Fibonacci numbers.

- **Daisies:** As mentioned earlier, daisies often have petal counts that are Fibonacci numbers, like 34, 55, or 89.

- **Roses:** Roses typically have petal counts that match Fibonacci numbers, with the most common varieties having 5, 8, 13, or 21 petals.

2. Pinecones: If you closely observe a pinecone, you'll notice that its scales are arranged in spiral patterns. These spirals curve both clockwise and counterclockwise around the cone. If you count the number of spirals going in one direction (say, clockwise) and then count the number going in the opposite direction (counterclockwise), these numbers are often consecutive Fibonacci numbers.

For example, in many pinecones, you might count 8 spirals going in one direction and 13 spirals going in the other. Both 8 and 13 are Fibonacci numbers. The Fibonacci sequence in pinecone spirals is not just a random occurrence. It happens because this arrangement allows the pinecone to pack its scales in the most efficient way possible. The spiral pattern helps maximize the space available for each scale, ensuring they fit tightly together as the cone grows. Similar to pinecones, the scales on a pineapple also exhibit Fibonacci spirals.

3. Shell Spirals:

- **Growth Patterns:** Many shells, such as those of snails and the chambered nautilus, grow in a spiral pattern. As the shell grows, it follows a logarithmic spiral, which means that the shape of the shell remains the same as it increases in size. The growth of these shells is proportional, and this proportional growth can be described by the Fibonacci sequence.

- **Logarithmic Spirals:** A logarithmic spiral is a type of spiral where the size of the spiral increases, but the shape remains constant. The growth factor in these spirals is often linked to the golden ratio. When you look at the cross-section of a shell, the spiral shape you see is a visual representation of the Fibonacci sequence in action.

- **Shell Structure and Fibonacci:** As the shell grows, each new section added to the spiral is proportional to the previous one, following the ratios found in the Fibonacci sequence. For example, if a particular section of the shell corresponds to a Fibonacci number, the next section will relate to the next number in the sequence.

- **Efficiency of Space:** The reason behind this growth pattern is efficiency. The Fibonacci sequence allows for the most efficient use of space, enabling the organism to grow without changing shape. The proportionality helps the shell maintain its strength and stability as it expands.

4. Hurricane Patterns: The spiral shape of hurricanes is governed by the Fibonacci sequence, showing how even chaotic natural events follow mathematical laws.

- **Spiral Shape**: Hurricanes often exhibit a spiral structure when viewed from above, resembling the spiral patterns seen in other natural forms like shells and galaxies. This spiral shape is not just random; it follows a logarithmic spiral, which is mathematically related to the Fibonacci sequence.

- **Logarithmic Spirals and Fibonacci**: The spiral arms of a hurricane curve outward as they move away from the center, forming a pattern similar to the Fibonacci spiral. In a logarithmic spiral, the spacing between the arms increases as the spiral winds outward, and this increase can be described using Fibonacci numbers.

- **Energy Efficiency and Stability**: The spiral pattern of a hurricane is an efficient way for the storm to distribute its energy. The Fibonacci sequence and golden ratio allow the hurricane to maintain stability as it grows in size and strength, ensuring that the wind and rain are distributed evenly across the storm.

Natural Examples:

Hurricane Katrina: When looking at satellite images of Hurricane Katrina, the spiral pattern of the storm can be mapped onto a Fibonacci spiral, illustrating the natural occurrence of this mathematical sequence in large-scale weather systems.

Typhoons and Cyclones: Other types of cyclones, including typhoons, also display this spiral pattern, adhering to the Fibonacci sequence and demonstrating the universal application of this mathematical concept in nature.

5. Human Hand: Interestingly, the Fibonacci sequence can also be observed in the structure and proportions of the human hand.

Finger Bones: If you look at the bones in your fingers, you'll notice that they follow a pattern that corresponds to the Fibonacci sequence. Starting from the tip of the finger, the bones increase in length as you move toward the base of the hand. Specifically:

- The bone in the tip of the finger (the distal phalanx) is approximately 1 unit long.

- The next bone (the middle phalanx) is roughly 2 units long.

- The bone at the base of the finger (the proximal phalanx) is about 3 units long.

These lengths approximate the Fibonacci sequence (1, 2, 3), showing how the structure of the hand adheres to this mathematical pattern.

Hand Structure: The human hand consists of five fingers, each with a different number of joints and segments. If you count the number of joints from the tip of each finger to the base (including the knuckles), you'll notice that they follow the Fibonacci pattern:

- The thumb has 2 joints.

- The index finger has 3 joints.

- The middle finger has 3 joints.

- The ring finger has 3 joints.

- The pinky has 3 joints.

These numbers (2, 3, 3, 3, 3) loosely correspond to the Fibonacci sequence and demonstrate how nature organizes living structures according to these mathematical principles.

Applications in Art and Design: The proportions of the human hand, guided by the Fibonacci sequence, have been used by artists and designers for centuries to create works of art that are visually harmonious. The idealized human hand, with its Fibonacci-based structure, has been a model for sculptures, paintings, and even ergonomic designs.

The Beauty of Golden Ratio

The **Golden Ratio** is a special number, approximately equal to 1.618, which has been used in art, architecture, and nature for centuries due to its aesthetically pleasing proportions. Mathematically, the Golden Ratio is often denoted by the Greek letter φ (phi).

If you have a line segment, the Golden Ratio occurs when you divide the line into two parts such that the ratio of the whole line to the longer part is the same as the ratio of the longer part to the shorter part.

Mathematically, if a line segment is divided into two parts, a and b, where a is the longer part and b is the shorter part, the Golden Ratio is defined by the equation:

$$\frac{a+b}{a} = \frac{a}{b} = \varphi = 1.618 \text{ (approximately)}$$

One of the most well-known examples of the Golden Ratio in the human body is the ratio between the length of your forearm and the length of your hand.

Start by measuring the distance from your elbow to your wrist. Let's say this distance is 26 cm (this will be different depending on the person).

Next, measure the distance from the base of your hand (where your wrist ends) to the tip of your middle finger. Let's say this distance is 16 cm.

Now, divide the length of your forearm by the length of your hand.

$$\text{Ratio} = \frac{Forearm\ Length}{Hand\ Length} = \frac{26}{16} \approx 1.625 \text{ (very close to the Golden Ratio)}$$

Other examples:

The Parthenon: The ancient Greek Parthenon was designed using the Golden Ratio, creating a visually pleasing and harmonious structure.

Mona Lisa: Leonardo da Vinci used the Golden Ratio to structure the composition of the Mona Lisa, giving it a balanced and aesthetically pleasing appearance.

Human Face: The proportions of the human face, such as the distance between the eyes and the length of the nose, often reflect the Golden Ratio, which is why it is associated with beauty.

DNA Molecule: The dimensions of the DNA double helix are in the ratio of 21 to 34 angstroms per turn, which are Fibonacci numbers that closely approximate the Golden Ratio.

Galaxy Spirals: The spiral arms of galaxies often follow the Golden Ratio, illustrating how this mathematical principle is embedded in the very fabric of the universe.

Fractals in Nature

Imagine looking at a tree. If you zoom in on one of its branches, you'll notice that the branch looks a lot like a smaller version of the whole tree. If you zoom in even further on a twig coming off that branch, it also resembles the original tree shape. This repeating pattern, where each smaller part looks similar to the whole, is what we call a fractal.

Fractals are patterns that repeat at different levels of scale. No matter how much you zoom in or zoom out, the pattern stays the same. These patterns are not just mathematical ideas; they appear all around us in nature. For example, the branching patterns of trees, the shape of coastlines, and even the veins in leaves are all examples of fractals. This repeating nature shows us that nature itself is built on these intricate and beautiful patterns that repeat over and over again.

- **Fern Leaves**: If you look closely at a fern leaf, you will see that each smaller leaf resembles the entire fern. This is a fractal pattern, where each part reflects the whole.

- **Coastlines**: The jagged edges of coastlines are fractals. No matter how much you zoom in, the pattern of the coastline remains irregular and complex.

- **Clouds**: The fluffy, uneven edges of clouds are another example of fractals. Each small part of a cloud looks similar to the whole.

- **Snowflakes**: Snowflakes have a fractal structure, with each arm of the snowflake resembling the others, creating a complex, symmetrical pattern.

- **Mountain Ranges**: The roughness of mountain ranges can be described using fractal geometry. Each peak and valley has a pattern that is self-similar across different scales.

Fractals are important because they show us that nature is built on simple, repetitive rules that create complex and beautiful patterns. This repetition and self-similarity suggest that there is an underlying order in the universe, which some interpret as a reflection of the divine.

Mathematics of the Human Body

The human body is a marvel of mathematical design. From the way our cells grow to the proportions of our bodies, math is everywhere in our anatomy.

- **The Fibonacci Sequence in the Body**: The bones in our hands and arms follow the Fibonacci sequence. Each section of the finger is roughly the sum of the two preceding sections, creating a harmonious and functional structure.

- **The Golden Ratio in Human Proportions**: The ratio of the length of our forearm to our hand is close to the Golden Ratio, reflecting the aesthetic and functional design of the human body.

- **The Structure of DNA**: The DNA molecule, the blueprint of life, has dimensions that are Fibonacci numbers, demonstrating that the very code of life follows mathematical principles.

- **Heartbeat Rhythms as a Sine Wave**: The rhythm of the human heartbeat can be modeled using a sine wave, a fundamental mathematical function. This regularity is crucial for maintaining life.

- **Symmetry in Human Body**: Our body exhibits bilateral symmetry, meaning it can be divided into mirrored halves. This symmetry is mathematically significant and is a fundamental aspect of biology.

Mathematical Thinking as a Spiritual Practice

Mathematics is not just about numbers and equations; it can also be a path to spiritual enlightenment. Here are some ways in which mathematical thinking can be seen as a spiritual practice:

- **Meditation through Problem-Solving**: Engaging deeply with a mathematical problem can be meditative. The focus required to solve complex problems can bring clarity and peace of mind.

- **Contemplation of Infinity**: Many mathematical concepts, like infinity, encourage us to think beyond the physical world. Contemplating infinity can be a way to connect with the idea of the eternal or the divine.

- **Pattern Recognition as Mindfulness**: Recognizing mathematical patterns requires mindfulness and attention to detail. This practice can help us become more aware of the patterns in our own lives and the world around us.

- **Mathematics as a Path to Wisdom**: Understanding mathematics can lead to greater wisdom and insight into the workings of the universe, much like spiritual teachings offer wisdom about the nature of existence.

Mathematics Proves the Universe is a Living Force

Mathematics doesn't just describe the universe; it reveals that the universe is dynamic, interconnected, and alive. Here's how:

- **Dynamic Systems and Chaos Theory**: Chaos theory shows that even simple mathematical systems can produce unpredictable and complex behavior, much like the natural world. For example, weather patterns, population dynamics, and even the stock market exhibit chaotic behavior, where small changes can have large effects. This unpredictability suggests that the universe is not a static machine but a living, evolving system.

  Example: The Butterfly Effect, a concept from chaos theory, states that a small change in one part of a system can lead to significant changes in another part. This idea can be seen in how ecosystems respond to changes in the environment or how small actions in our lives can lead to significant consequences.

- **Fractals in Nature**: Fractals are complex patterns that are self-similar across different scales. They appear in natural phenomena such as the branching of trees, the structure of snowflakes, and the coastline of a landmass. The fact that fractals are found in so many aspects of nature suggests that the universe is built on patterns that are both intricate and self-organizing, characteristics of a living system.

- **Quantum Entanglement**: Quantum entanglement is a phenomenon where particles become linked, such that the state of one particle is instantly connected to the state of another, no matter the distance between them. This interconnectedness at the quantum level suggests that all parts of the universe are in constant communication, supporting the idea that the universe operates as a unified, living entity.

- **Mathematical Models of Life**: Mathematical models, such as those used in biology to describe population growth, neural networks, and the spread of diseases, show that life processes are deeply mathematical. These models help us understand that life itself is governed by mathematical laws, further blurring the line between mathematics and the essence of living systems.

  Example: The mathematical modeling of how neurons in the brain fire in patterns, creating thoughts and consciousness, shows the mathematical foundation of living processes.

The Essential Role of Mathematics in the Work of World-Renowned Scientists

Mathematics is often considered the language of science, and many of the most famous scientists in history were also accomplished mathematicians. Their groundbreaking discoveries were deeply rooted in mathematical principles, demonstrating that understanding and applying mathematics is essential for scientific progress. Here's how mathematics played a crucial role in the work of some of the world's most renowned scientists:

1. Isaac Newton (1643–1727)

- **Scientific Contributions:** Newton formulated the laws of motion and universal gravitation, which became the foundation of classical mechanics.

- **Mathematical Contributions:** To solve problems in physics, Newton invented calculus (independently of Leibniz). His mathematical innovations were essential in expressing the laws of motion and gravitation, proving that mathematics is key to understanding physical laws.

2. Albert Einstein (1879–1955)

- **Scientific Contributions:** Einstein's theory of relativity revolutionized our understanding of space, time, and gravity.

- **Mathematical Contributions:** Einstein employed complex mathematical concepts like tensors and non-Euclidean geometry to develop his theories. His famous equation $E=mc^2$ mathematically links energy and mass, underscoring the role of mathematics in deciphering the mysteries of the universe.

3. James Clerk Maxwell (1831–1879)

- **Scientific Contributions:** Maxwell's equations describe electromagnetism and unified electricity, magnetism, and light as interconnected phenomena.

- **Mathematical Contributions:** Maxwell's work relied on a set of partial differential equations, demonstrating that mathematical equations are fundamental in describing the behavior of electric and magnetic fields.

4. Galileo Galilei (1564–1642)

- **Scientific Contributions:** Often called the "father of modern science," Galileo made significant advances in physics, astronomy, and the scientific method.

- **Mathematical Contributions:** Galileo used mathematical analysis to study motion, laying the groundwork for Newton's laws. His application of geometry and arithmetic to physical problems showed that mathematics is vital to scientific discovery.

5. Leonhard Euler (1707–1783)

- **Scientific Contributions:** Euler made major contributions to mechanics, fluid dynamics, and astronomy.

- **Mathematical Contributions:** Euler is celebrated as one of history's greatest mathematicians, with work in calculus, number theory, and topology. His mathematical insights were foundational in the advancement of modern science, highlighting the integral role of mathematics in scientific innovation.

6. Niels Bohr (1885–1962)

- **Scientific Contributions:** Bohr developed the Bohr model of the atom, introducing the concept of quantized energy levels.

- **Mathematical Contributions:** The Bohr model is based on mathematical equations that describe electron behavior in atoms. Bohr's reliance on quantum mechanics, a highly mathematical field, underscores the connection between mathematics and scientific progress.

7. Richard Feynman (1918–1988)

- **Scientific Contributions:** Feynman was instrumental in developing quantum electrodynamics (QED), which explains how light and matter interact.

- **Mathematical Contributions:** Feynman's work in QED involved advanced mathematical tools, such as path integrals and Feynman diagrams, showing that sophisticated mathematics is necessary to understand and predict physical phenomena.

8. Marie Curie (1867–1934)

- **Scientific Contributions:** Curie's research on radioactivity led to the discovery of polonium and radium.

- **Mathematical Contributions:** Although known primarily for her experimental work, Curie used mathematics to analyze her data, quantify radioactivity, and understand radioactive decay. Her work demonstrates that mathematics is crucial for interpreting experimental results in science.

These examples illustrate that many of the world's greatest scientists were also deeply engaged with mathematics. Their scientific breakthroughs were not just supported by mathematics but were often driven by it. This highlights the essential role that mathematical understanding plays in the development of scientific theories, the analysis of data, and the making of accurate predictions about the natural world.

How Math Can Keep Your Mind Young and Healthy as You Age

Learning math can be a powerful tool to keep your brain healthy and active, especially as you get older. Here's how math can contribute to a longer, healthier life:

1. **Keeps Your Brain Engaged:** Math exercises your brain, just like physical exercise strengthens your body. Regularly engaging in math problems helps maintain mental sharpness, reducing the risk of cognitive decline as you age.

2. **Strengthens Memory:** Practicing math challenges your brain to remember formulas, methods, and steps, which can improve your overall memory. This mental workout helps your brain stay strong, aiding in better recall and memory retention.

3. **Enhances Problem-Solving Skills:** Math teaches you to think logically and solve problems methodically. These skills are valuable in everyday situations, helping you navigate challenges more effectively, which in turn can reduce mental stress.

4. **Reduces Stress and Anxiety:** Successfully solving math problems can provide a sense of accomplishment and boost your confidence. This positive experience can lower stress levels, which is important for maintaining good health as you age.

5. **Promotes Brain Flexibility:** Learning new math concepts encourages your brain to form new neural connections, keeping it flexible and adaptable. This mental agility is crucial for staying mentally sharp as you grow older.

While math alone doesn't directly extend your lifespan, its positive impact on brain health can contribute to a more active, engaged, and fulfilling life as you age.

Conclusion

Mathematics is more than just a subject; it's a way of connecting with the deepest truths of the universe. By learning math, we are not just learning about numbers and equations—we are learning about the very fabric of reality.

This understanding can bring us closer to the divine, helping us see the order, beauty, and infinite complexity of the world around us.

In this way, math is not just something we study in school but a lifelong journey of discovery—a path to understanding the universe and a bridge to connecting with God. By embracing math, we embrace the very principles that govern existence, deepening our connection to the divine.

Story Time: Meera's Journey from Chaos to Clarity through Mathematics

Introduction

Meera was a math genius from a young age. She understood numbers, patterns, and equations easily, and she excelled in all her math subjects in school and college. Her logical thinking and problem-solving skills made her stand out, and she was often praised for how easily she solved complex math problems. But as Meera grew older, she realized that life's problems weren't as straightforward as the math problems she used to solve in textbooks.

Even though Meera was talented in math, she often felt overwhelmed by the challenges in her personal life. Whether it was managing her money, dealing with relationships, or organizing her daily tasks, she struggled. This constant state of chaos left her feeling anxious, unfocused, and unable to move forward with confidence. She began to feel that her math skills were not useful in the messy, unpredictable world of adulthood.

Problem: Chaos and Lack of Direction

Meera's life was disorganized, and she lacked clarity. This chaos affected her personal, professional, and financial life, leaving her feeling stuck and unsure of how to move forward. No matter how hard she tried, she found it difficult to bring order to her life and make meaningful progress.

Solution: Using Mathematical Thinking to Bring Order

One day, Meera attended a workshop that explained how mathematical thinking could be used in everyday life. The idea that math could help bring clarity and order to life resonated deeply with her. The workshop showed how the principles of math could be applied to solve everyday problems, helping to create structure and improve decision-making. For Meera, this was a revelation;

it was as if the math mindset she had developed in school could now be used to solve the chaos in her life.

Understanding the Power of Mathematics

Meera realized that the logical, structured thinking she had mastered in school could be applied to her personal life. She learned that math wasn't just about solving equations but also about recognizing patterns, understanding relationships, and breaking down complex problems into smaller, manageable parts. This new understanding reignited her passion for math and gave her tools to tackle the chaos in her life.

Applying Mathematical Logic to Finances

Meera started by applying mathematical logic to her finances. She treated her finances like an equation, where each expense and income was a variable to be managed. This systematic approach allowed her to gain control over her money, helping her see where she was overspending and where she could save.

Example: Meera created a financial plan where she listed all her income and expenses, like balancing an equation. By identifying and cutting unnecessary expenses, she gradually solved her financial problems.

Using Geometry to Organize Her Space

Meera also applied geometric principles to her living space. She understood that geometry is about understanding shapes, spaces, and how they fit together. By organizing her home using geometric ideas—like symmetry and proportion—she created a more harmonious and efficient living environment. This not only made her space more functional but also brought a sense of calm and order to her daily life.

Example: Meera rearranged her furniture using principles of symmetry, creating a balanced and pleasing layout. This made her feel more at ease in her home and reduced the stress of clutter.

Implementing Statistical Thinking in Relationships

Meera began to apply statistical thinking to her relationships. By observing patterns in her interactions and analyzing past experiences, she identified behaviors that led to conflicts and those that strengthened her relationships. This data-driven approach allowed her to make better decisions in her social interactions, leading to healthier and more fulfilling relationships.

Example: Meera kept a journal where she noted key interactions with friends and family. Over time, she used this information to identify patterns—like noticing that certain topics always led to arguments with her sibling. This helped her avoid conflicts and build stronger connections.

Using Algebra to Set and Achieve Goals

Meera applied algebraic thinking to her life goals. Just as algebra involves solving for an unknown variable by breaking down an equation, Meera broke down her life goals into smaller, manageable tasks. This method made her goals seem less overwhelming and more achievable.

Example: Meera wanted to save for a vacation. She created a "goal equation" where her savings target was the solution. She then identified the variables—cutting unnecessary expenses, increasing her income, and setting aside a specific amount each month. This approach allowed her to reach her savings goal without feeling overwhelmed.

Developing Resilience through Calculus

Meera realized that calculus, which focuses on change and growth, offered valuable lessons in resilience. Life, like a calculus function, is about continuous change—sometimes growing, sometimes declining. By embracing this mindset, Meera became more adaptable and better equipped to handle life's ups and downs.

Example: When faced with a challenging project at work, Meera approached it like a calculus problem, breaking it down into small steps. This allowed her to build momentum and steadily progress toward completion, even when the task seemed overwhelming.

Conclusion

By applying mathematical thinking, Meera was able to bring clarity and order to her life. The mathematical mindset she had developed in school became a valuable tool in navigating the complexities of adulthood. Math gave her the tools to analyze her problems, break them down into manageable parts, and find solutions that brought structure and balance. Meera not only improved her life but also contributed to a broader sense of positivity and growth.

Key Insight

Mathematical thinking offers a powerful way to solve life's problems by providing clarity, structure, and logical strategies. Meera's story shows how the math skills she developed in her youth became a vital tool in her adult life, helping her bring order to chaos, solve complex problems, and achieve her goals. This approach emphasizes the importance of structured, logical thinking in creating a balanced and harmonious life.

CHAPTER 18

Religious God vs. Super-Consciousness

Introduction

"The universe is not made of atoms; it's made of tiny stories."

– Muriel Rukeyser

All around the world, people turn to religion for comfort, guidance, and answers to life's biggest questions. The idea of God varies from one culture and belief system to another, but it usually involves a higher power that rules over the universe. However, what if we could think about God differently—not as a distant, all-powerful being, but as a universal consciousness that connects everything? This chapter will explore how the concept of God could be reinterpreted as super-consciousness, a force we can understand and interact with through science, rather than through mysticism.

We'll use principles from game theory—a way to understand strategic decision-making—to explain how our individual minds are part of a larger, interconnected super-consciousness. This super-consciousness can be seen as your personal "god," not in the traditional sense of worship, but as a scientific understanding of how we fit into the universe. In this view, becoming part of this consciousness is more powerful than simply worshipping a deity.

Religious God vs. Super-Consciousness

Understanding God in Traditional Religion

In most religions, God is seen as an all-powerful being who controls the universe and intervenes in human lives. God is worshipped, prayed to, and asked for guidance and protection. This God is often thought of as separate from the physical world, existing in a realm beyond human understanding. People rely on faith to connect with this God and seek answers to the mysteries of life.

177

Reinterpreting God as Super-Consciousness

Super-consciousness suggests that all individual minds are connected in a universal field. This field is not separate from us; it is part of reality itself. Instead of being a distant deity, super-consciousness is a force that connects all living things. It operates not through magic or miracles, but through the natural laws of the universe.

Super-consciousness is the source of creativity, intuition, and the deep connections that link us to the universe and each other. It exists within us and around us, guiding us not through religious teachings but through our experiences, thoughts, and actions.

Game Theory and the Connection Between Consciousness and Super-Consciousness

1. What is Game Theory?

Game theory is a way to understand how people make decisions when their choices affect others. Imagine you and a friend are deciding what game to play. If you both want to play the same game, you'll have fun. But if you want different things, you'll need to figure out how to cooperate or compromise. Game theory helps us understand how to make decisions that benefit everyone involved.

2. The Prisoner's Dilemma: A Simple Example

Let's look at the Prisoner's Dilemma. Imagine two friends are caught doing something wrong and are taken to separate rooms. Each friend is given a choice: If one confesses and the other doesn't, the one who confesses goes free, and the other gets a long punishment. If both confess, they both get a medium punishment. But if neither confesses, they both get a light punishment.

The best outcome for both friends would be if they trusted each other and didn't confess. But because they can't talk to each other and might not trust each other completely, they often both end up confessing, getting a worse outcome than if they had cooperated.

How This Relates to Super-Consciousness:

Think of your mind as one of the friends and super-consciousness (the collective mind of all people) as the other. When you make decisions that only benefit you, it can lead to negative consequences for everyone, including yourself. But if you make decisions that consider the well-being of others and the world (the super-consciousness), everyone benefits more.

For example, if you help someone in need, even if it's a small sacrifice for you, it strengthens the bonds in your community, making life better for everyone, including you.

3. Nash Equilibrium: Finding Balance

Another idea in game theory is called the Nash Equilibrium. This is a situation where everyone involved in a decision has found a strategy that works for them, and no one can do better by changing their choice. It's like when you and your friends all agree on a game to play that everyone enjoys, so no one has a reason to suggest changing the game.

How This Relates to Super-Consciousness:

In life, we're constantly making choices that affect not just ourselves but everyone around us. If we all try to make choices that benefit both ourselves and others, we can reach a state where everything is in balance—where everyone is happy with their decisions and how they affect the larger community.

For example, think of a workplace where everyone is doing a job that suits their skills and interests. Everyone is satisfied because they're doing something they're good at and enjoy, and the company as a whole benefits. This balance in the workplace is like a Nash Equilibrium.

Applying Game Theory to Everyday Life

Here's how understanding these simple game theory concepts can help you in everyday life:

1. **Cooperation Over Competition:** When you face a choice, think about how your decision will affect others. Cooperating with others often leads to better outcomes for everyone, just like in the Prisoner's Dilemma when neither friend confesses.

Example: If you're working on a group project, instead of focusing on getting credit just for yourself, focus on how the group can succeed together. This cooperation will likely lead to a better project and a better experience for everyone.

2. **Seeking Balance:** Try to find a balance in your choices that benefits both you and those around you. This is like finding a Nash Equilibrium in your life.

Example: If you're trying to balance time between your family and work, find a schedule that allows you to succeed at your job while also spending quality time with your loved ones. This balanced approach benefits both your career and your personal life.

Game theory shows us that our decisions are interconnected, just like how our minds are connected to the super-consciousness. By making choices that benefit not just ourselves but also others, we contribute to a more balanced, harmonious world. This idea is at the heart of super-consciousness—understanding that we are all part of a larger whole and that our actions can shape reality.

The Science of Becoming God

Traditional religion often places God on a pedestal, separate from humanity, as an entity to be worshipped and revered. However, the concept of super-consciousness offers a different perspective—one where each person has the potential to embody god-like qualities by understanding and connecting with the universal consciousness.

Becoming a Part of Super-Consciousness

Becoming God in this context means recognizing our role as integral parts of the super-consciousness. By developing our self-awareness—through practices like meditation, self-reflection, and compassionate action—we contribute to the strength of the super-consciousness. This isn't about gaining supernatural powers but about realizing our potential to positively influence the world and create harmony in the universe.

The Role of Science in Understanding God

Science, particularly in fields like quantum mechanics and game theory, offers tools to understand the universe in ways that traditional religion doesn't. While religious texts provide moral and ethical guidance, science reveals the underlying mechanisms that govern reality. Super-consciousness, seen through the lens of science, becomes a unifying force that goes beyond religious boundaries. It helps us see that the divine isn't something outside of us but something that exists within us and throughout the universe.

Super-Consciousness as a Unifying Force

Super-consciousness can be viewed as the ultimate unifying force that transcends individual religions and beliefs. It doesn't demand worship or blind faith but encourages understanding, growth, and connection. This concept of God is inclusive, scientific, and rooted in our interconnected existence. It invites us to see the divine not as a distant figure but as the collective consciousness that we are all part of, encouraging us to contribute positively to the whole.

How Religions Reflect Game Theory Principles for a Balanced Life

Every major religion in the world incorporates principles that align with game theory, a concept from mathematics that explores strategies for cooperation, competition, and achieving balanced outcomes. These religious teachings emphasize the importance of living harmoniously, making choices that benefit not only oneself but also others. Let's explore how different religions reflect these principles:

1. Christianity: The Golden Rule

Christianity teaches the Golden Rule: "Do unto others as you would have them do unto you." This simple yet profound teaching emphasizes the importance of treating others with the same kindness and respect that you would like to receive. In game theory, this concept reflects the strategy of cooperation, where individuals consider the well-being of others to achieve better outcomes for everyone. By fostering mutual respect and empathy, the Golden Rule encourages a society where people work together, creating a more harmonious and just world.

2. Islam: The Concept of Ummah

In Islam, the concept of *Ummah* refers to the global Muslim community, where all members are connected through shared faith and mutual responsibilities. The teachings of *Zakat* (charity) and social justice emphasize the importance of sharing wealth and resources to support those in need. This collective responsibility mirrors game theory's focus on cooperative strategies, where the well-being of the community as a whole is prioritized. By encouraging acts of charity and justice, Islam promotes a balanced society where resources are distributed more equitably, ensuring that everyone benefits.

3. Hinduism: The Law of Karma

Hinduism teaches the principle of *Karma*, which highlights the cause-and-effect relationship between actions and their outcomes. According to this belief, every action has consequences, and individuals are encouraged to act in ways that are beneficial not only to themselves but also to others. This aligns with game theory, where the choices one makes can influence the overall outcome of the "game" of life. By promoting actions that lead to positive outcomes for both the individual and the community, Hinduism fosters a balanced life, where good deeds are rewarded, and harmful actions are discouraged.

4. Buddhism: The Principle of Right Action

Buddhism emphasizes the principle of *Right Action*, which encourages ethical conduct and compassion towards others. This principle is part of the Noble Eightfold Path, a guide to living a life that leads to enlightenment. The emphasis on interdependence—understanding that all beings are connected—aligns closely with game theory. In game theory, the choices one person makes can have ripple effects, influencing the larger system. By teaching that one's actions affect the well-being of the entire community, Buddhism encourages individuals to act in ways that contribute to the collective good, promoting peace and harmony.

5. Judaism: The Concept of Tikkun Olam

In Judaism, the concept of *Tikkun Olam*, or "repairing the world," is a call to action for Jews to contribute to the betterment of society. This concept is rooted in the ethical teachings of justice, kindness, and fairness. By encouraging individuals to work towards the common good, *Tikkun Olam* reflects game theory principles where cooperative strategies lead to better outcomes for all.

Whether it's through acts of charity, social justice, or simply treating others with respect, Judaism teaches that everyone has a role in creating a more just and balanced world.

6. Sikhism: The Practice of Seva

Sikhism emphasizes the importance of *Seva*, or selfless service, as a way to live in harmony with others. This practice is seen as a way to connect with the divine and contribute to the well-being of the community. In game theory, cooperative strategies often lead to mutually beneficial outcomes. By encouraging Sikhs to engage in acts of service without expecting anything in return, Sikhism promotes a society where individuals work together for the greater good, leading to a more harmonious and balanced life.

7. Taoism: Harmony with the Tao

Taoism focuses on living in harmony with the *Tao*, which represents the natural order of the universe. The concept of *Wu Wei*, or effortless action, teaches that by aligning oneself with the flow of nature, one can achieve balance and peace without forcing outcomes. This idea mirrors game theory's principle of finding equilibrium, where the best results are often achieved not by aggressive competition but by going with the flow and making choices that harmonize with the environment. Taoism encourages individuals to live simply, in tune with the rhythms of life, promoting a balanced existence where everything falls into place naturally.

Across all these religions, the underlying message is clear: living a balanced life requires cooperation, ethical behavior, and consideration for others. These principles, reflected in religious teachings, are also central to game theory, where the focus is on strategies that benefit both the individual and the community. By following these teachings, individuals can contribute to a more harmonious and just world, where everyone has the opportunity to thrive.

Conclusion

The traditional concept of God is deeply rooted in human culture and spirituality. However, as we learn more about the universe through science, we are offered a new perspective—one where God is not a separate entity to be worshipped but a super-consciousness that we are all part of. By embracing this idea, we can move beyond traditional religious thought and realize our potential as creators of our reality.

Through game theory, we understand that our individual actions contribute to the super-consciousness, shaping our lives and the universe itself. Becoming God means recognizing our role in the greater whole, using our consciousness to create, sustain, and nurture the world around us. This approach aligns with the goals of super-consciousness and unites us all in a shared quest for understanding, growth, and harmony.

Story Time: Nisha's Journey from Daily Boredom to Living with Purpose

Introduction

Nisha was a hardworking and responsible person who seemed to have an ideal life. She had a stable job, a loving family, and a comfortable home. But despite these blessings, Nisha often felt bored and uninspired by the predictability of her daily routine. Every day felt the same—waking up, going to work, coming home, cooking dinner, and then going to bed. This repetitive pattern started to make her feel tired and uninterested in life. She often daydreamed about vacations or exciting events that might bring some excitement to her life.

However, even when these events happened, the excitement didn't last long, and Nisha quickly found herself back in the same cycle of boredom. She began to wonder why life felt so repetitive and why her days seemed to pass by without any real meaning. It felt like she was just going through the motions, not truly living.

Problem: Boredom and Lack of Purpose

Nisha's life was overshadowed by boredom and dissatisfaction with her repetitive routine. She felt disconnected from the present, always waiting for something exciting to happen that would bring meaning to her life. This constant longing for change left her feeling unfulfilled and stuck in a cycle of routine.

Solution: Embracing Super-Consciousness through Mathematics and Game Theory

One day, Nisha read an article about the connection between consciousness, super-consciousness, and the logical structure of life explained through mathematics and game theory. The idea that life could be understood and made meaningful through math and strategy intrigued her. She decided to explore this further, hoping it would help her find new purpose in her daily life.

Rediscovering Purpose through Mathematical Thinking

Nisha started by exploring how the structured thinking found in mathematics could apply to her life. She realized that just as in math, where every problem has a solution and every equation has a purpose, her daily routine could be given meaning and purpose. She began to see each task, no matter how small, as part of a larger plan. By applying problem-solving principles from math, she started to approach her daily activities with a new mindset.

Example: Nisha turned meal planning into a fun challenge, trying to balance nutrition and budget like solving a math problem. Each day became an opportunity to improve her strategies, making her routine feel more engaging.

Applying Game Theory to Daily Decisions

Nisha also learned about game theory, which is usually used in economics or politics but can also be applied to personal life. Game theory helped her understand that life is full of decisions, each with its own possible outcomes. By viewing her choices through this lens, she saw that even the most mundane decisions could be approached strategically.

Example: Instead of dreading her commute, Nisha used it as an opportunity to listen to educational podcasts that aligned with her goals. This turned her commute into a productive part of her day, making her feel more in control of her life.

Connecting with Super-Consciousness

Through her exploration of mathematics and game theory, Nisha began to feel a deeper connection with the concept of super-consciousness. She realized that her individual decisions, no matter how small, were part of a larger, interconnected system. This understanding gave her actions a new sense of purpose. Nisha started to see her life as part of something bigger, where every choice she made contributed to the well-being of herself and those around her. This made her feel more connected to the world and less isolated in her routine.

Transforming Routine into Mindful Practice

With this new understanding, Nisha transformed her daily routine into a mindful practice. She approached each day with intention, seeing her activities not as repetitive chores but as actions that contributed to her overall goals. Whether at work, with her family, or during self-care, Nisha applied mindfulness to stay fully present. This shift in perspective helped her find joy in the simplicity of her daily tasks, turning the mundane into something meaningful.

Aligning with the Science of Becoming

Nisha also embraced the idea that super-consciousness was not something outside of her, but a part of her being. She learned that by becoming more aware of her thoughts and actions, she could align with this higher consciousness. Instead of waiting for something external to bring her happiness, she began to create fulfillment from within. This practice of self-awareness and intentional living helped Nisha break free from boredom and led her to a more enriched and purposeful life.

Sharing Her Journey

Inspired by her transformation, Nisha began sharing her journey with others who felt trapped in their routines. She encouraged them to explore the principles of mathematics and game theory as tools for living more intentionally. Nisha's story resonated with many, showing that the key to breaking free from monotony and finding fulfillment lies in how we approach our daily lives, not in waiting for something extraordinary to happen.

Key Insight

Nisha's story shows the power of applying mathematical thinking and game theory to everyday life. By viewing her routine through the lens of logic and strategy, she transformed her daily tasks from boring chores into purposeful actions. This approach brought new meaning to her life and connected her to the broader concept of super-consciousness. Nisha's journey demonstrates that fulfillment and purpose come from within and that by aligning our actions with higher principles, we can find joy and meaning even in the simplest aspects of life.

CHAPTER 19

Non-Duality in Hinduism

Introduction

For centuries, the concept of non-duality has been a central idea in Hindu philosophy. Known as *Advaita*, meaning "not two," this belief suggests that there is no separation between the self (*Atman*) and the ultimate reality (*Brahman*). In simpler terms, it means that everything in the universe is interconnected, and what we see as separate things are actually different parts of the same underlying reality.

But what if this ancient idea isn't just spiritual poetry? What if it has a scientific basis in one of the most advanced fields of modern physics—Quantum Mechanics? In this chapter, we'll explore how the principles of non-duality in Hinduism align with ideas in Quantum Mechanics and String Theory, offering a new way to understand the interconnectedness of everything.

The Essence of Non-Duality

To understand non-duality, think of the ocean. The ocean is vast and deep, and it creates countless waves. Each wave looks like a separate thing, but in reality, all waves are just expressions of the ocean. They are not independent; they are connected to the ocean and to each other. This is the essence of non-duality: the idea that everything we see, experience, and are is part of a single, unified reality.

In Hinduism, *Brahman* is like the ocean—the ultimate reality—and everything in the universe, including ourselves, is like a wave on that ocean. The waves seem separate, but they are not different from the ocean. Similarly, we may think of ourselves as separate individuals, but according to non-duality, we are all connected to each other and to the universe in a deep, meaningful way.

Quantum Mechanics: The Science of Interconnection

Now, let's talk about Quantum Mechanics, a branch of physics that studies how particles behave at the smallest scales. One of the most fascinating ideas in Quantum Mechanics is *quantum entanglement*. When two particles become entangled, the state of one particle is directly linked to the state of the other, no matter how far apart they are. If you change something about one particle, the other instantly reflects that change, even if they are on opposite sides of the universe.

This challenges our usual understanding of separation and distance. In the quantum world, the boundaries we assume between objects don't really exist. Everything is interconnected in ways that go beyond our everyday experience. This is where the connection to non-duality becomes clear: just as Quantum Mechanics shows that particles are not truly separate, non-duality suggests that we are not truly separate from each other or from the universe.

The Role of Consciousness

Another interesting concept in Quantum Mechanics is the role of the observer in determining the state of a quantum system. Before an observation is made, particles exist in a state of probability, known as a superposition. It's the act of observing that causes these probabilities to collapse into a single state. This idea fits well with the non-dualistic view that consciousness plays a key role in shaping reality.

In Hindu philosophy, *Atman*—the self—is not just an individual consciousness but a part of the universal consciousness (*Brahman*). This consciousness is not passive; it actively participates in creating reality. Similarly, in Quantum Mechanics, the observer is not separate from the observed; the act of observing helps bring reality into existence. This connection between the observer and reality in both Quantum Mechanics and non-duality highlights the deep link between consciousness and the universe.

The Bhagavad Gita and Modern Science: A Meeting of Minds

The Bhagavad Gita, a 700-verse Hindu scripture that is part of the Indian epic Mahabharata, is a conversation between Prince Arjuna and the god Krishna, who serves as his charioteer. This ancient text is not just a spiritual guide but also a profound exploration of the nature of reality, consciousness, and the self.

In the Gita, Krishna teaches Arjuna about the soul (*Atman*), which is eternal and beyond physical existence. He explains that this soul is not born, nor does it die; it goes beyond the physical body and is one with the ultimate reality, *Brahman*. This teaching reflects the core of non-dualistic philosophy, emphasizing the unity of the self with the universe.

Modern science, especially Quantum Mechanics, offers a parallel to this concept. The idea that particles can exist in a state of superposition—where they are not in one state or another until observed—can be seen as a reflection of the Gita's teaching on the impermanent nature of physical reality. Just as the soul in the Gita transcends the physical body, quantum particles exist beyond fixed states until consciousness interacts with them.

Example: Schrödinger's Cat and the Soul's Eternal Nature

In Quantum Mechanics, there is a famous thought experiment called *Schrödinger's Cat*, which illustrates the concept of superposition. Imagine a cat inside a sealed box, along with a mechanism that has a 50% chance of releasing poison and killing the cat, and a 50% chance of leaving the cat unharmed. According to quantum theory, until we open the box and observe the cat, it exists in a superposition of states—both alive and dead at the same time. It's only when we observe the cat that it "chooses" a definite state, either alive or dead.

This idea of superposition shows that at the quantum level, reality is not fixed until it is observed. The physical state of the cat is uncertain, existing in multiple possibilities until consciousness (through observation) collapses it into one reality.

Now, let's relate this to the Bhagavad Gita's teaching about the soul (*Atman*). In the Gita, Lord Krishna explains to Arjuna that the soul is eternal and indestructible—it is beyond birth and death, and it transcends the physical body. The physical body is like the box in Schrödinger's thought experiment, while the soul is like the cat. Just as the cat's state is not fixed until observed, the physical body is impermanent and ever-changing, but the soul remains constant and unbound by these changes.

In essence, the soul in the Gita represents a state that goes beyond the temporary, physical existence (the body), much like how quantum particles transcend fixed states until they are observed. The Gita teaches that the physical world is fleeting and subject to change, but the soul—the true self—is eternal and unchanging. Similarly, Quantum Mechanics shows that at the most basic level, reality is not fixed or permanent until observed, aligning with the idea that our material existence is temporary, while our true essence is something beyond physical form.

This comparison helps us understand that both the Gita and modern science suggest that what we perceive as reality is not as solid or fixed as it appears. In both views, there is a deeper, underlying truth that goes beyond the physical world: in the Gita, it's the eternal soul; in Quantum Mechanics, it's the uncertain quantum state that becomes fixed only through observation.

Real-World Examples of Interconnection

To make these abstract concepts more understandable, let's look at some real-world examples that show the principle of interconnection:

1. **The Butterfly Effect**: In chaos theory, the Butterfly Effect describes how small changes in one part of a system can lead to significant effects in another part. For instance, the flapping of a butterfly's wings in Brazil could theoretically set off a tornado in Texas. This concept mirrors the idea of non-duality, where even the smallest action can influence the entire universe.

2. **Ecosystems**: In nature, ecosystems show the principle of interconnection beautifully. Every organism in an ecosystem is linked to others in a delicate balance. A change in one species can have ripple effects throughout the entire system. This interdependence is a real-world reflection of the non-dualistic view that nothing exists in isolation.

3. **Global Economy**: The global economy is another example of how interconnected our world is. A financial crisis in one country can trigger economic repercussions around the globe. This interconnectedness, where local actions have global consequences, aligns with the quantum view that everything is part of a larger, interconnected whole.

The Living Universe: Reflections from Indian Mythology

Indian mythology is rich with stories that portray the universe as a living, conscious entity, deeply interconnected with all beings. These tales suggest that the cosmos is not just a physical space but a dynamic, breathing organism infused with divine energy. Below are ten stories that beautifully illustrate this concept, offering profound insights into the nature of the universe.

1. The Churning of the Ocean (Samudra Manthan)

Reference: Mahabharata and various Puranas

The gods and demons churn the cosmic ocean to obtain the nectar of immortality, symbolizing the universe's life-giving potential. The emergence of divine beings and objects from the ocean represents the living, dynamic nature of the universe, capable of creating and sustaining life.

2. The Cosmic Dance of Shiva (Nataraja)

Reference: Shiva Purana and other texts

Lord Shiva's dance of creation, preservation, and destruction reflects the cyclical nature of the universe. This dance, known as the Tandava, illustrates the universe as a living entity in perpetual motion, where every movement signifies the rhythms of life and the cosmos.

3. Vishnu's Universal Form (Vishvarupa)

Reference: Bhagavad Gita, Chapter 11

In the Bhagavad Gita, Lord Krishna reveals his Vishvarupa, a form that embodies the entire universe. This vision shows the universe as a living organism, where all beings and elements are interconnected within the divine body of Vishnu, symbolizing the unity and consciousness of the cosmos.

4. The Birth of the Devas from Aditi

Reference: Rigveda and various Puranas

Aditi, the mother of the gods, prayed to the universe for children who would protect it. Her prayers were answered, and she gave birth to the Devas, who represent natural forces. This story underscores the universe as a nurturing, living force that responds to and sustains life.

5. The Dream of Vishnu

Reference: Vishnu Purana and other texts

The universe is depicted as a dream of Lord Vishnu, who lies on the cosmic serpent Ananta. As Vishnu dreams, the universe comes into existence and dissolves when he awakens. This narrative emphasizes the universe as a living, conscious manifestation of divine thought.

6. The Birth of Brahma from Vishnu's Navel

Reference: Bhagavata Purana and Vishnu Purana

Brahma, the creator god, emerged from a lotus that grew out of Lord Vishnu's navel, symbolizing the universe as a living organism. Vishnu represents the sustaining force, and Brahma represents creative energy, both of which highlight the interconnectedness of the cosmos.

7. The Descent of the Ganges (Ganga Avatarn)

Reference: Ramayana and various Puranas

The Ganges is personified as a goddess whose descent to Earth symbolizes the flow of divine energy into the world. This story portrays the universe as a living entity that nurtures life through sacred elements like rivers, which are seen as the lifeblood of the cosmos.

8. The Story of Varaha and the Earth (Bhudevi)

Reference: Vishnu Purana and other texts

When the demon Hiranyaksha dragged the Earth (personified as Bhudevi) to the bottom of the cosmic ocean, Lord Vishnu, as Varaha, rescued her. This tale highlights the Earth as a living being, emphasizing the universe's vitality and the divine care it receives.

9. Markandeya and the Cosmic Deluge

Reference: Markandeya Purana

Sage Markandeya witnessed a cosmic deluge where the universe was submerged in water. He saw a tiny child, representing the living universe, floating on a banyan leaf. This child was Lord Krishna, symbolizing the eternal life force within the cosmos.

10. The Formation of Shakti Peethas

Reference: Shiva Purana and Devi Bhagavata Purana

After Sati's self-immolation, her body parts fell to Earth, creating sacred sites known as Shakti Peethas. This story reflects the universe as a living entity where divine energy (Shakti) is present in all creation, making the entire cosmos sacred and alive.

These stories from Indian mythology provide a profound perspective on the universe as a living, conscious entity. They suggest that the cosmos is a dynamic, interconnected organism, infused with divine consciousness and responsive to the energies of all beings. These narratives remind us of the sacredness of the universe and our intrinsic connection to it.

Conclusion: The Unity of Science and Spirituality

As we've seen, the ancient philosophy of non-duality in Hinduism and the modern science of Quantum Mechanics share a common idea: that everything in the universe is interconnected. This connection is not just a metaphor—it's a fundamental aspect of reality that both science and spirituality recognize.

The Bhagavad Gita, with its teachings on the eternal nature of the soul, the interconnectedness of actions, and the role of consciousness, provides a spiritual foundation that finds a remarkable parallel in the principles of modern science. By understanding these connections, we can start to see the universe in a new way. We are not isolated beings living in a fragmented world; we are part of a vast, unified whole. Our thoughts, actions, and consciousness are all threads in the fabric of the universe, woven together by the force of Universal Consciousness.

In this chapter, we've only scratched the surface of this profound idea. As we continue our journey in the next chapter, we'll explore how other Eastern philosophies, like Taoism and Buddhism, also resonate with the principles of modern physics, offering even more insights into the true nature of reality.

CHAPTER 20

Eastern Wisdom and String Theory

Introduction

"The essence of the universe is vibration.
All existence is interconnected and flows as one harmonious melody."

In the previous chapter, we explored how Hinduism's concept of non-duality aligns with Quantum Mechanics. Now, let's take this exploration further by looking at how other Eastern philosophies, like Taoism and Buddhism, connect with one of the most advanced theories in modern physics—String Theory.

String Theory suggests that the basic building blocks of the universe are not tiny particles, but small, vibrating strings of energy. These strings vibrate at different frequencies to create everything in the universe—from the smallest particles to the largest cosmic structures. Interestingly, this scientific idea of a connected, unified reality is similar to the teachings of Taoism and Buddhism, which also emphasize the oneness of existence and the flow of energy through all things.

The Core of String Theory

To understand how String Theory relates to Eastern philosophies, we first need to grasp its basic idea. Traditional physics views particles like electrons and quarks as tiny points with no size or structure. However, String Theory suggests that if we could zoom in close enough, we would see that these particles are actually tiny loops of vibrating strings.

These strings can vibrate in different ways, and each vibration corresponds to a different type of particle. For example, one vibration might create an electron, while another might create a photon. The various particles and forces in the universe come from the different ways these strings vibrate.

Example:

Think of a guitar string. When you pluck it, the string vibrates to produce a musical note. The note you hear depends on the vibration's frequency. Similarly, in String Theory, the type of particle—whether it's an electron, quark, or photon—depends on how the string vibrates. Different vibrations create different particles, just like different notes from a guitar string. This suggests that everything in the universe is interconnected through these strings, which vibrate together to form the fabric of reality.

String Theory also introduces the idea of extra dimensions. While we experience the world in three spatial dimensions and one time dimension, String Theory suggests there could be more dimensions beyond our perception. These extra dimensions might be hidden but still influence how strings vibrate, affecting the physical properties of our universe.

Example:

Imagine a garden hose lying on the ground. From far away, the hose looks like a simple line (one dimension). But when you get closer, you see that it has a circular cross-section—an extra dimension. Similarly, the extra dimensions in String Theory might be hidden from our view but still play a crucial role in determining how strings vibrate and how the universe behaves.

Taoism: The Flow of the Tao

Taoism is an ancient Chinese philosophy that focuses on the concept of the Tao, often translated as "the Way." The Tao is the underlying force that flows through everything, connecting and harmonizing the universe. It is both the source and the flow of all existence, present in everything but beyond full understanding or naming.

The Tao is like a river, constantly moving and adapting to the world around it. This flow is not chaotic; it follows a natural course, guiding everything toward balance and harmony.

Example:

Imagine a river flowing through a valley. The river represents the Tao—always moving, shaping everything in the valley, from rocks to trees. Similarly, the Tao influences everything in the universe. This idea of flow and balance is like the vibrations in String Theory, where strings interact to create the physical world. Just as the Tao represents the flow of energy and the connectedness of all things, String Theory suggests that the universe is made up of vibrating strings that connect and influence each other.

Taoism also emphasizes Wu Wei, which means "non-action" or going with the flow of the Tao rather than resisting it. This concept can be compared to the natural vibration of strings in String Theory, where strings create reality effortlessly. The principle of Wu Wei reflects the natural, harmonious vibration of strings, creating the universe without force or resistance.

Example:

Think about learning to play a musical instrument. At first, you might struggle to play even simple notes. But as you practice, it becomes easier, and you can play effortlessly, without thinking about each movement. This is similar to Wu Wei—acting in harmony with the Tao, where your actions become natural and spontaneous. In the same way, the vibrations of strings in String Theory happen naturally, creating the universe without any external force.

Buddhism: Interconnectedness and the Middle Way

Buddhism is another ancient Eastern philosophy that teaches the principle of interconnectedness through the concept of Pratītyasamutpāda—the idea that everything arises in dependence on other things. In other words, everything in the universe is connected, and nothing exists on its own. This idea matches the interconnected nature of strings in String Theory, where the vibration of one string affects the vibrations of others, creating a web of interdependence that forms the fabric of reality.

Example:

Imagine a vast network of dominoes set up in a complex pattern. When you knock over one domino, it causes a chain reaction, toppling others in sequence. Now, imagine these dominoes are interconnected in such a way that knocking over one can set off multiple chains of reactions throughout the network.

This interconnectedness is similar to the Buddhist idea that all things are dependent on each other. It also mirrors String Theory's concept that the universe is made up of interconnected strings. Just as tipping one domino can create a cascade of effects throughout the network, a single vibration in one string can influence the entire cosmos, showing the deep interconnectedness of all things.

Buddhism also teaches the Middle Way, a path of moderation that avoids extremes. This idea of balance and harmony is similar to how strings vibrate in String Theory—each string must be in perfect balance to create the particles and forces that make up our universe. Just as the Middle Way guides one toward enlightenment by avoiding extremes, the balanced vibrations of strings guide the creation of reality.

Buddhism also teaches that the self is an illusion—what we see as the self is just a collection of temporary, interconnected processes. This idea aligns with String Theory, where matter is not solid or fixed but a dynamic, interconnected web of vibrating strings. The Buddhist concept of Śūnyatā, or emptiness, which teaches that all things are empty of inherent existence, mirrors the String Theory idea that what we see as solid matter is really made of vibrating energy.

Example:

Think of a whirlpool in a river. The whirlpool looks like a separate thing, but it's just a pattern of water movement created by the river's flow. When the flow changes, the whirlpool disappears, showing that it was never separate from the river. Similarly, Buddhism teaches that the self is a temporary pattern within the flow of existence—an illusion of separateness created by interconnected processes. In String Theory, matter is not solid or fixed but a dynamic pattern of vibrating strings, showing that what we see as solid and separate is actually part of a larger, interconnected whole.

Real-World Applications: The Resonance of Eastern Wisdom in Modern Science

To bring these ideas into the real world, let's look at some examples where Eastern philosophy and modern science intersect:

1. **Meditation and Brain Waves:** Meditation, a practice central to both Taoism and Buddhism, has been shown to change brain waves, harmonizing the mind and body. This is a real-world example of how

tuning into the flow of the Tao or the interconnectedness of all things can create harmony within oneself, just like how the harmonious vibrations of strings create a balanced universe.

2. **Holistic Medicine:** Eastern philosophies often emphasize treating the body and mind as a whole, recognizing that all aspects of health are connected. This holistic approach is gaining recognition in modern medicine, where there's growing awareness of the connection between mental, emotional, and physical health. This reflects the idea in String Theory that everything is connected and that small changes can have widespread effects.

Example:

In holistic medicine, treating an illness involves not just addressing physical symptoms but also considering the mental and emotional well-being of the patient. For example, acupuncture, an ancient practice rooted in Eastern philosophy, is believed to balance the flow of energy within the body, restoring harmony and promoting healing. This idea aligns with String Theory's concept that everything is connected, and balancing the vibrations of strings can lead to a harmonious and healthy existence.

3. **Environmental Sustainability:** The interconnectedness of all things is a central theme in both Eastern philosophy and String Theory. This idea has practical implications for how we treat the environment. Understanding that our actions have far-reaching consequences can inspire more sustainable practices, reflecting the Taoist and Buddhist teachings of living in harmony with nature.

Example:

Think about deforestation's impact on the global ecosystem. Cutting down trees not only affects the local environment but also contributes to global climate change, disrupts wildlife habitats, and influences weather patterns. This interconnectedness of environmental factors mirrors the idea in String Theory that every vibration affects the whole. Just as one string's vibration can influence the entire universe, our actions can have significant effects on the environment, highlighting the importance of living in harmony with nature.

The Living Universe: Reflections from Western Mythology and Folklore

Western mythology and folklore also offer numerous stories that suggest the universe is not just a physical space, but a living, breathing entity interconnected with all forms of life. These tales often depict the cosmos as a dynamic, conscious force that interacts with humans and other beings, influencing their destinies and the world around them. Here are ten stories from the Western world that illustrate this concept:

1. Gaia, the Earth Mother

Reference: Greek Mythology

Gaia is the personification of the Earth in Greek mythology, depicted as a living entity who gave birth to the sky (Uranus), the mountains, and the sea. Gaia is the primordial mother who nurtures all life, symbolizing the Earth itself as a living, conscious being that sustains all existence.

2. The Norse World Tree (Yggdrasil)

Reference: Norse Mythology

Yggdrasil, the immense sacred tree, connects the nine worlds in Norse mythology. It is considered a living entity that holds the cosmos together, with its roots and branches stretching into different realms of existence. The health of Yggdrasil directly affects the state of the universe, symbolizing the interconnectedness and vitality of the cosmos.

3. The Myth of Prometheus

Reference: Greek Mythology

Prometheus is the Titan who stole fire from the gods and gave it to humanity, symbolizing the gift of knowledge and enlightenment. This myth reflects the idea that the universe is alive with knowledge and energy, which can be accessed and utilized by living beings, showing a deep connection between cosmic forces and human existence.

4. The Celestial Sphere

Reference: Ancient Greek and Medieval Cosmology

The concept of the celestial sphere in ancient and medieval thought portrays the universe as a series of concentric, rotating spheres with Earth at the center.

These spheres were believed to be alive, with each one governed by a divine being or force that controlled the movement of stars and planets, suggesting a universe imbued with life and intelligence.

5. The Tale of Pandora's Box

Reference: Greek Mythology

Pandora's box, which contained all the evils of the world, also released hope when it was opened. This story reflects the dual nature of the universe as a living entity that holds both positive and negative forces. The presence of hope among the evils symbolizes the universe's capacity to nurture and sustain life even amidst chaos.

6. The Creation of the World by the Titans

Reference: Greek Mythology

In Greek mythology, the Titans were the children of Gaia (Earth) and Uranus (Sky), and they played a crucial role in shaping the universe. Their actions led to the formation of the cosmos as we know it, emphasizing the universe as a dynamic, living force shaped by powerful beings.

7. The Roman Genius Loci

Reference: Roman Mythology

Genius Loci refers to the protective spirit of a place in Roman religion and mythology. Every location, whether it be a grove, river, or mountain, was believed to have its own spirit, reflecting the idea that the Earth and its places are alive and conscious, each with its own personality and protective qualities.

8. The Phoenix Rising from the Ashes

Reference: Greek and Roman Mythology

The Phoenix, a mythological bird that is reborn from its ashes, symbolizes the cycle of life, death, and rebirth within the universe. This story illustrates the idea of the cosmos as a living, regenerating entity that constantly renews itself, reflecting the eternal nature of life in the universe.

9. The Legend of the Green Man

Reference: Celtic and British Folklore

The Green Man is a figure symbolizing nature's rebirth and growth, often depicted as a face surrounded by or made from leaves. He represents the living Earth and the life force that permeates all vegetation, embodying the idea of the natural world as a conscious, life-giving entity.

10. The Cosmic Serpent (Ouroboros)

Reference: Ancient Egyptian and Greek Symbolism

The Ouroboros, a serpent eating its own tail, represents the eternal cycle of life, death, and rebirth. This symbol is found in various Western traditions, depicting the universe as a self-sustaining, living entity that is continually renewing itself. It suggests a cosmos that is both alive and cyclical, reflecting the continuous process of creation and destruction.

These stories from Western mythology and folklore offer a perspective that the universe is not just a collection of inert matter but a living, conscious entity. Through these narratives, the cosmos is portrayed as dynamic, interconnected, and infused with a life force that touches all beings within it. Whether through the nurturing Earth Mother Gaia, the cosmic cycles of the Phoenix, or the life-sustaining World Tree Yggdrasil, these tales remind us of the sacred and living nature of the universe.

Conclusion: The Harmony of Science and Philosophy

As we've seen, the principles of String Theory resonate deeply with the wisdom of Eastern philosophies like Taoism and Buddhism. Both suggest that the universe is not a collection of separate objects, but a harmonious, interconnected whole where everything is balanced.

String Theory's idea that the universe is made up of vibrating strings echoes the Taoist concept of the flowing Tao and the Buddhist principle of interconnectedness. These ancient philosophies, like String Theory, encourage us to see beyond the illusion of separation and recognize the unity of all things.

By understanding these connections, we can see how ancient wisdom and modern science are not in conflict but are two ways of understanding the same truth: that everything in the universe is interconnected, and by living in harmony with this reality, we can find balance and peace in our lives.

CHAPTER 21

Answering Life's Toughest Questions

Introduction

"The important thing is not to stop questioning. Curiosity has its own reason for existing."

– Albert Einstein

Throughout our lives, we often face difficult questions that seem impossible to answer, especially when we look to religion for explanations. These questions touch on deep issues like suffering, natural disasters, the purpose of life, and the existence of evil. Sometimes, traditional religious answers might feel unsatisfying or incomplete. In this chapter, we'll explore how the concept of Super-Consciousness—a universal, intelligent force that connects everything—can offer more logical and fulfilling answers to these tough questions.

1. Why Do Bad Things Happen to Good People?

Super-Consciousness Perspective:

The suffering of good people is one of the most challenging aspects of life to understand. From the perspective of Super-Consciousness, the experiences we call "bad" are not punishments but opportunities for growth. These experiences help individuals develop qualities like resilience, empathy, and wisdom. Every challenge, no matter how painful, helps expand consciousness, pushing people to explore their inner strengths and overcome their limitations.

Example:

Consider Viktor Frankl, an Austrian psychiatrist who survived the Holocaust. Despite losing almost everything, he found meaning in his suffering, which led him to develop "logotherapy"—a therapy focused on finding meaning in life. His experiences show that even in the darkest times, there is an opportunity to find purpose and grow.

For someone under stress or in need, understanding that their suffering has a purpose—however hidden it may be—can bring comfort. It allows them to see their pain as part of a larger, transformative journey that can lead to personal and spiritual growth.

2. Why Does God Allow Natural Disasters?

Super-Consciousness Perspective:

Natural disasters are often seen as random and devastating events. From a Super-Consciousness perspective, these disasters are not acts of cruelty but part of the Earth's ecological balance. They remind us of life's fragility and the importance of unity, compassion, and resilience in the face of adversity.

Example:

The 2011 earthquake and tsunami in Japan caused immense suffering, but it also brought out extraordinary acts of bravery and kindness. The disaster led to changes in nuclear energy policies and disaster preparedness, showing how such events can prompt important lessons and positive changes.

For those affected by disasters, it can be hard to find meaning in the chaos. Yet, from a Super-Consciousness perspective, these events offer opportunities for humanity to come together and learn valuable lessons that can lead to a safer, more connected world.

3. What Happens After We Die?

Super-Consciousness Perspective:

Death is often feared as the ultimate end. However, from the perspective of Super-Consciousness, death is not an end but a transformation. The consciousness that makes us who we are does not disappear; it transitions into a new state, merging with the greater universal consciousness. The experiences and growth achieved in life continue to influence the collective consciousness of the universe.

Example:

The Tibetan Book of the Dead describes death as a passage through different states of consciousness, leading to rebirth or liberation. This view suggests that death is not the end but a transition to another phase of existence.

For someone afraid of death, understanding it as a transition rather than an end can be comforting. It reassures them that their life's experiences and relationships are not lost but become part of a greater, eternal consciousness.

4. Why Are There So Many Religions?

Super-Consciousness Perspective:

The existence of many religions raises questions about truth and the nature of the divine. From the perspective of Super-Consciousness, the diversity of religions reflects the diversity of human cultures and experiences. Each religion offers a different path to understanding the same fundamental truths, adapted to the needs of different societies. Instead of conflicting, these religions are complementary, each adding to humanity's collective wisdom.

Example:

Think of religions as different languages. Just as different languages express the same ideas in unique ways, religions express universal truths through diverse teachings and symbols. The Bahá'í Faith, for example, teaches that all religions are part of a single, progressive revelation from God, emphasizing that diversity in religious expression reflects the richness of the human experience.

For someone struggling with religious doubt or confusion, this perspective can be liberating. It encourages respect for all spiritual paths and appreciation for the universal truths they convey.

5. Does God Answer Prayers?

Super-Consciousness Perspective:

Prayer is a deeply personal practice that transcends religious boundaries. From the Super-Consciousness perspective, prayer is not about asking for specific outcomes but about aligning oneself with the universal consciousness. When prayers seem to be answered, it's often because the individual's intentions have harmonized with the natural flow of the universe. When prayers go unanswered, it might be because the request does not align with the individual's true path or the broader needs of the universe.

Example:

Consider Corrie ten Boom, a Dutch Christian who helped Jews escape the Nazis during World War II. Despite her prayers, she and her family were caught and sent to concentration camps, where many of her loved ones died.

Ten Boom later reflected that her experience deepened her faith, even though her prayers were not answered as she had hoped.

For someone feeling disillusioned because their prayers seem unanswered, this perspective encourages them to see prayer as a way of aligning with life's deeper currents, trusting that a higher wisdom is guiding their journey.

6. Why Is God Invisible?

Super-Consciousness Perspective:

God's invisibility is often seen as a challenge to faith. From the perspective of Super-Consciousness, God is not a separate, visible entity but the very fabric of the universe itself. This divine presence is felt through experiences, intuitions, and inner awareness rather than seen with physical eyes.

Example:

Think of the wind. We cannot see the wind, but we can feel it and see its effects. Similarly, we cannot see God, but we can feel the divine presence in moments of deep peace, love, and connection. The story of Elijah in the Bible, where God is found in a "still small voice," highlights how the divine is experienced in quiet, inner moments rather than grand, visible displays.

For someone struggling with doubt because they cannot see God, this perspective offers reassurance. It invites them to cultivate inner awareness and trust in the divine presence that guides and supports them.

7. Is There Free Will, or Is Everything Predestined?

Super-Consciousness Perspective:

The debate over free will versus predestination is ancient. From the Super-Consciousness perspective, both free will and predestination coexist. Certain life events may be influenced by the flow of Super-Consciousness, but individuals exercise free will in how they respond to these events. Free will allows consciousness to explore and expand within the framework of the universe's design.

Example:

Imagine a river flowing towards the sea. The river's course is shaped by the landscape (predestination), but within this course, the river is free to flow in various ways (free will). Similarly, while certain aspects of our lives are beyond our control, we still have the power to shape our journey through our choices.

For someone struggling with the tension between free will and destiny, this perspective offers a harmonious resolution, acknowledging that we can shape our lives even within certain limitations.

8. Why Does God Create Life Only to Let It Die?

Super-Consciousness Perspective:

The cycle of life and death is fundamental to existence. From the perspective of Super-Consciousness, life and death are not opposites but parts of a continuous process of transformation. Death is not an end but a transition where individual consciousness evolves and merges with the broader Super-Consciousness.

Example:

Consider the life cycle of a tree. A tree grows, lives, and eventually dies, but its death contributes to new life in various forms, such as providing nourishment to the soil. Similarly, human life and death are part of a larger cycle of existence.

For someone grieving the loss of a loved one, this perspective offers comfort by framing death as a natural and meaningful part of life's journey.

9. Why Are Some People Born into Suffering and Others into Privilege?

Super-Consciousness Perspective:

The circumstances of birth—whether into suffering or privilege—offer specific opportunities for growth. Each soul enters life with conditions that best support the experiences it needs for its evolution, helping to develop qualities like compassion, resilience, or humility.

Example:

Oprah Winfrey was born into poverty and suffered abuse as a child, yet she used these experiences to grow into a compassionate and influential leader. Her hardships shaped her ability to connect with others and advocate for change.

For someone struggling with feelings of injustice, this perspective offers a way to reframe their experiences as unique opportunities for growth tailored to their soul's journey.

10. Why Are Miracles Rare?

Super-Consciousness Perspective:

Miracles are rare because they represent moments of extraordinary alignment between individual or collective consciousness and Super-Consciousness. These events occur when the natural flow of the universe aligns with deep intention and awareness, producing outcomes that seem to defy normal expectations.

Example:

John Smith, a teenager who fell through ice and was submerged in freezing water for 15 minutes, miraculously survived despite being without a pulse for 45 minutes. This event was widely regarded as a miracle, reflecting a deeper alignment between the faith of his loved ones and the life force within him.

For someone yearning for a miracle, understanding its rarity can be both humbling and hopeful, encouraging patience and trust in the process of life.

11. What Is the Purpose of Life?

Super-Consciousness Perspective:

The purpose of life is to contribute to the ongoing expansion and evolution of consciousness. Life is a journey of exploration, learning, and growth, where each experience serves to refine consciousness and move it closer to alignment with Super-Consciousness.

Example:

Nelson Mandela's life, marked by imprisonment and suffering, ultimately led to his role in ending apartheid and promoting justice and reconciliation. His life's purpose was not about personal gain but contributing to the greater good.

For someone searching for meaning, this perspective encourages them to consider how their life can contribute to the greater evolution of consciousness.

12. Why Doesn't God Speak to Us Directly?

Super-Consciousness Perspective:

From the perspective of Super-Consciousness, God is not a separate entity that communicates in human language. Instead, the divine presence is felt through experiences, intuitions, and inner realizations. God communicates through synchronicities, inspirations, and the inner voice that guides us.

Example:

Joan of Arc claimed to hear divine voices guiding her to lead France to victory. Whether these voices were external or internal, her life demonstrates how profound inner guidance can feel like divine communication.

For someone feeling abandoned or unheard by God, this perspective suggests that divine communication is not about hearing words but about being attuned to the subtle, inner guidance that is always present.

13. Why Does Evil Exist in the World?

Super-Consciousness Perspective:

Evil exists as part of the duality that defines human experience. It provides the contrast necessary for the growth of consciousness. By overcoming evil, individuals and societies develop greater strength, wisdom, and compassion, contributing to the overall evolution of consciousness.

Example:

The Holocaust was one of the most horrific examples of evil in history, but it also led to countless acts of courage and compassion, showing that in the presence of great evil, great good can also emerge.

For someone struggling to understand why evil exists, this perspective offers a way to see it not as senseless suffering but as part of the larger process of spiritual evolution.

14. Why Are Innocent Children Born with Disabilities or Terminal Illnesses?

Super-Consciousness Perspective:

The birth of innocent children with disabilities or terminal illnesses is one of life's most heart-wrenching mysteries. From the perspective of Super-Consciousness, these conditions offer unique opportunities for growth in compassion, resilience, and unconditional love.

Example:

Nick Vujicic, born without arms and legs, became a motivational speaker, inspiring millions with his message of hope and perseverance. His life shows that even the most challenging circumstances can lead to profound growth and impact.

For someone grappling with the pain of seeing a child suffer, this perspective offers a way to find meaning in the midst of heartache.

15. Why Are Some Prayers Unanswered Despite Strong Faith?

Super-Consciousness Perspective:

Unanswered prayers can be deeply disheartening, especially when one has placed all their faith in a specific outcome. From the perspective of Super-Consciousness, prayers may go unanswered when the request does not align with the broader flow of the universe or the individual's path of growth.

Example:

Joni Eareckson Tada, who became a quadriplegic after a diving accident, prayed for healing but remained paralyzed. Over time, she found new purpose in life, becoming an artist, author, and advocate for people with disabilities. Her story illustrates that unanswered prayers can lead to unexpected growth and alignment with a higher purpose.

For someone feeling disillusioned because their prayers seem unanswered, this perspective offers a new way to understand their experience, encouraging them to trust in the broader flow of Super-Consciousness.

Conclusion: Embracing the Journey of Consciousness

As we navigate life's complexities, we often face questions that challenge our understanding. Why do good people suffer? Why does evil exist? What is the purpose of life? These questions reflect our deep need to find meaning amid uncertainty and pain.

Through the lens of Super-Consciousness, we can see these challenges not as obstacles but as essential components of our journey toward greater awareness and spiritual evolution. Life's difficulties, whether personal or global, are opportunities for growth, both individually and collectively.

Each experience, no matter how painful or confusing, serves to expand our consciousness.

This perspective encourages us to embrace life's dualities—joy and sorrow, light and darkness, life and death—not as opposites but as complementary aspects of the same profound truth. It teaches us that suffering and adversity are not punishments but invitations to grow, deepen our empathy, and contribute to the collective evolution of consciousness.

As we move forward, let us carry with us the understanding that we are part of a vast, interconnected web of existence. Our lives, with all their challenges and triumphs, are threads in the intricate tapestry of Super-Consciousness. In embracing this journey, we discover that the true purpose of life is not to find all the answers but to live the questions fully, with an open heart and a courageous spirit.

CHAPTER 22

Connecting the Dots

As we approach the final stages of our journey, it's time to weave together the diverse threads we've explored so far. We've delved into the intricate connections between science and spirituality, examining how modern physics, ancient wisdom, and the concept of Super-Consciousness intersect. Now, we'll bring these ideas into focus, connecting them to form a cohesive understanding of how they collectively shape a new vision of the universe and our role within it. This chapter serves as a bridge, helping us see the bigger picture and how each concept we've discussed contributes to a deeper, more integrated understanding of existence.

Looking Back at the Journey

We started by questioning the traditional idea of God, exploring how ancient religious teachings and modern scientific theories might not just coexist but actually complement each other. We moved away from the image of God as a distant being and toward a more abstract but deeply connected idea of Super-Consciousness. This Super-Consciousness isn't something separate from us; it's the very essence of the universe, the source from which everything flows.

We also examined concepts like karma and the flow of energy in the universe, how our thoughts could be seen as quantum states, and how consciousness might be a fundamental part of everything that exists. Through these explorations, we've found a new way to think about spirituality—not as something opposed to science, but as an integral part of it.

Bringing Science and Spirituality Together

One of the most important insights we've gained is that science and spirituality are not in conflict. Instead, they offer different but complementary ways of understanding reality. Quantum Mechanics, for instance, teaches us about the interconnectedness of particles and the role of the observer in shaping reality. This idea mirrors spiritual teachings that emphasize the unity of all things and the power of our consciousness.

String Theory, with its idea that the universe is made up of tiny, vibrating strings, aligns with the ancient belief that the universe is a living, interconnected whole. These scientific theories provide a framework for understanding the universe that resonates deeply with spiritual insights, giving us a new perspective on age-old questions about existence, purpose, and the divine.

What This Means for Us

So, what does all of this mean in our everyday lives? It means that our thoughts, actions, and beliefs are powerful—they are the building blocks of reality. By understanding the connections between our consciousness and the universe, we can start to take more responsibility for the world we create—both in our personal lives and in the broader world.

This new perspective encourages us to live more mindfully, recognizing that everything we think and do contributes to the larger whole. It asks us to engage with life not just as passive observers but as active participants in the unfolding of reality. By aligning our consciousness with the principles of Super-Consciousness, we can live more meaningful, purposeful lives, contributing to the evolution of the universe itself.

Seeing the Big Picture

As we connect the dots between science and spirituality, we begin to see the universe in a new way. This vision doesn't ask us to give up our beliefs or choose between science and faith. Instead, it invites us to expand our understanding, to see the bigger picture where both science and spirituality are threads woven into the same tapestry of existence.

By embracing this holistic view, we begin to understand that the universe is not a cold, mechanical system but a living, breathing entity filled with consciousness and intelligence. This perspective shifts our view of life from one of randomness and separation to one of purpose and connection. We start to see that every event, every experience, and every interaction is part of a larger, interconnected whole—guided by the principles of Super-Consciousness.

Practical Implications

What are the practical implications of this new understanding? First, it changes how we relate to ourselves and others. When we recognize that we are all part of the same universal consciousness, it becomes easier to practice compassion, empathy, and love. We begin to see others not as separate from us but as extensions of the same consciousness that flows through us. This understanding can transform our relationships, helping us to connect more deeply with those around us.

Second, it changes how we approach challenges and difficulties in life. Instead of seeing obstacles as random or meaningless, we can view them as opportunities for growth and learning. Each challenge we face is a chance to expand our consciousness and contribute to the evolution of the universe. This perspective empowers us to face difficulties with courage and resilience, knowing that they serve a greater purpose.

Third, this understanding encourages us to take care of the environment and the world we live in. If everything is connected, then our actions have far-reaching consequences. By living in harmony with nature and being mindful of our impact on the planet, we contribute to the health and well-being of the entire universe.

Moving Forward

As we conclude this chapter, it's important to remember that the journey doesn't end here. The ideas we've explored are not just theoretical concepts to be understood—they are principles to be lived. The real value of this knowledge lies in how we apply it in our daily lives.

Moving forward, I encourage you to continue exploring these ideas, to question, to reflect, and to integrate them into your life. Let this new understanding guide your actions, shape your beliefs, and inspire you to live in alignment with the Super-Consciousness that connects us all.

Final Thoughts

In connecting the dots between science and spirituality, we've uncovered a new way of seeing the world—one that is filled with meaning, purpose, and profound connection. This vision invites us to transcend the limitations of traditional thinking and embrace a more expansive, integrated view of reality.

By living in alignment with this understanding, we become co-creators of the universe, contributing to the ongoing evolution of consciousness. This is not just a philosophical idea—it's a call to action, a call to live with greater awareness, intentionality, and compassion.

As you close this book, I hope you feel inspired to look at the world with fresh eyes, to see the divine in the everyday, and to recognize your role in the vast, interconnected web of existence. The journey of discovery doesn't end here—it's just the beginning. Let's continue to explore, to learn, and to connect the dots as we move forward on this incredible journey called life.

CHAPTER 23

Now Take Charge of Your Life

As we reach the final chapter of this journey, it's time to bring everything we've explored together into a practical, empowering message: now, take charge of your life. The insights we've uncovered about the universe, consciousness, and the connection between science and spirituality are not just intellectual exercises—they are tools that can transform how you live.

Understanding Your Role in the Universe

One of the most profound realizations we've discussed is the idea that you are an integral part of the universe. You are not just a passive observer; you are an active participant in the unfolding of reality. Your thoughts, actions, and intentions ripple through the fabric of existence, influencing not just your life but the world around you. This understanding is powerful because it shifts your perspective from being a victim of circumstances to being a creator of your reality.

Setting Intentions with Clarity

Taking charge of your life begins with setting clear, intentional goals. But these goals should not just be about achieving material success or external validation. Instead, they should align with the deeper purpose of expanding your consciousness and contributing positively to the world. When your goals are aligned with this higher purpose, you'll find that the universe supports your efforts in ways you might never have imagined.

Practical Step: Take some time to reflect on what truly matters to you. What are the goals that resonate with your deepest values and aspirations? Write them down, and make a plan to start working toward them today.

Embracing Responsibility

With great power comes great responsibility. As you take charge of your life, it's important to recognize the responsibility that comes with this power.

This means being mindful of your actions and their impact on others and the environment. It means living with integrity, honesty, and compassion, knowing that your choices contribute to the collective consciousness of the universe.

Practical Step: Start by making small changes in your daily life. Be more mindful of how you interact with others, how you spend your time, and how you treat the environment. Each positive change you make adds up, creating a ripple effect that can transform your life and the world around you.

Overcoming Fear and Doubt

One of the biggest obstacles to taking charge of your life is fear. Fear of failure, fear of the unknown, and fear of stepping outside your comfort zone can all hold you back. But remember, fear is just a natural response to stepping into new territory. It's not a sign that you should turn back—it's a sign that you're on the right path.

Practical Step: When fear or doubt arises, don't ignore it or let it control you. Instead, acknowledge it, understand where it's coming from, and then take action anyway. The more you face your fears, the less power they will have over you.

Cultivating Resilience

Life will inevitably throw challenges your way. But as we've discussed, these challenges are not meant to defeat you—they are opportunities for growth. Cultivating resilience means developing the ability to bounce back from setbacks, to learn from failures, and to keep moving forward, no matter what.

Practical Step: Practice resilience by reframing challenges as opportunities. When you encounter a difficulty, ask yourself, "What can I learn from this? How can this experience help me grow?" This mindset shift can make all the difference in how you navigate life's ups and downs.

Staying Connected to the Bigger Picture

As you take charge of your life, it's important to stay connected to the bigger picture. Remember that your life is part of something much larger—a vast, interconnected universe filled with consciousness and intelligence. By staying connected to this bigger picture, you'll find a sense of purpose and direction that can guide you through even the most challenging times.

Practical Step: Develop a regular practice that helps you stay connected to the bigger picture. This could be meditation, spending time in nature, journaling, or simply taking a few moments each day to reflect on your place in the universe.

The Power of Community

While taking charge of your life is a personal journey, it doesn't have to be a solitary one. Surround yourself with a supportive community of like-minded individuals who share your values and aspirations. This community can provide encouragement, accountability, and inspiration as you work toward your goals.

Practical Step: Seek out communities, both online and offline, that align with your interests and values. Engage with others who are on a similar path, and don't be afraid to ask for support or offer it to others.

Creating a Life of Meaning and Purpose

Ultimately, taking charge of your life is about creating a life of meaning and purpose. It's about aligning your actions with your values, contributing to the greater good, and living in a way that is true to who you are. When you live with purpose, you'll find that life becomes more fulfilling, more joyful, and more rewarding.

Practical Step: Reflect on what gives your life meaning and purpose. What are the activities, relationships, and pursuits that make you feel most alive? Make a commitment to prioritize these things in your life, and let them guide your decisions and actions.

Conclusion: The Journey Ahead

As you move forward from this book, remember that the journey of taking charge of your life is ongoing. It's not about reaching a final destination but about continuously growing, learning, and evolving. Embrace the journey with an open heart and a curious mind, and know that you have the power to create the life you desire.

Now, it's time to take everything you've learned and put it into action. The universe is ready to support you, and the possibilities are limitless. Take charge of your life, and watch as the world transforms around you.

Frequently Asked Questions about Super-Consciousness and Spirituality

In this bonus chapter, we'll address some of the most common questions people have about Super-Consciousness and spirituality, especially in relation to the ideas we've explored throughout this book. These questions often arise when people begin to contemplate the deeper aspects of existence, the nature of reality, and the connections between science and spirituality. My hope is that these answers will provide further clarity and help you integrate these concepts into your own life.

1. What exactly is Super-Consciousness?

Answer: Super-Consciousness is the idea that there is a universal, intelligent force that connects everything in the universe. It goes beyond individual consciousness and encompasses the collective awareness of all beings. This concept suggests that we are all interconnected at a fundamental level, and that our thoughts, actions, and experiences contribute to and are influenced by this greater consciousness. It's a way of understanding the universe as a living, dynamic entity where everything is linked.

2. How does Super-Consciousness relate to traditional concepts of God?

Answer: Super-Consciousness can be seen as a redefinition of God, one that aligns with modern scientific understanding. Instead of viewing God as a separate, anthropomorphic being who governs the universe from a distance, Super-Consciousness suggests that the divine is the very fabric of the universe itself. It's an intelligent force that permeates all things, guiding the evolution of consciousness and the unfolding of the universe. This concept is compatible with the idea that God is not separate from us but is an integral part of our existence.

3. Can Super-Consciousness be scientifically proven?

Answer: Super-Consciousness, like many spiritual concepts, is challenging to prove using traditional scientific methods because it deals with aspects of reality that transcend physical measurement. However, scientific theories like

Quantum Mechanics and String Theory suggest that the universe operates in ways that are consistent with the idea of an interconnected, conscious reality. While direct proof may be elusive, these theories provide a framework that supports the possibility of Super-Consciousness.

4. How can I connect with Super-Consciousness in my daily life?

Answer: Connecting with Super-Consciousness involves cultivating awareness and mindfulness in your daily activities. Practices like meditation, reflection, and being present in the moment can help you tune into this deeper level of consciousness. It's also important to approach life with an open mind, recognizing the interconnectedness of all things and the impact of your thoughts and actions on the broader universe. By fostering a sense of unity with the world around you, you can strengthen your connection to Super-Consciousness.

5. How does Super-Consciousness influence our lives?

Answer: Super-Consciousness influences our lives by guiding the flow of energy, thoughts, and experiences that shape our reality. When we align our personal consciousness with this greater force, we can experience a greater sense of purpose, clarity, and harmony in our lives. It helps us to see beyond the surface of everyday experiences and recognize the deeper connections that bind us to the universe and to each other.

6. What role does spirituality play in understanding Super-Consciousness?

Answer: Spirituality is the lens through which we explore and understand Super-Consciousness. It involves a journey of self-discovery, where we seek to understand our place in the universe and our connection to the divine. Spirituality encourages us to look beyond the material world and explore the deeper aspects of existence, leading us to a greater understanding of Super-Consciousness and how it influences our lives.

7. Is there a difference between individual consciousness and Super-Consciousness?

Answer: Yes, there is a difference. Individual consciousness refers to the awareness and experiences of a single person. It is the mind that perceives, thinks, and makes decisions on a personal level. Super-Consciousness, on the other hand, is the collective consciousness that encompasses all individual consciousnesses. It's the greater awareness that connects all beings and transcends the limitations of individual perception, providing a broader perspective on reality.

8. How does the concept of karma fit into the idea of Super-Consciousness?

Answer: Karma, the principle that every action has consequences, fits naturally into the concept of Super-Consciousness. Since Super-Consciousness is an interconnected force that links all beings, the effects of our actions reverberate throughout this network. Positive actions contribute to harmony and growth within the Super-Consciousness, while negative actions create disturbances. Understanding karma through the lens of Super-Consciousness helps us see how our choices affect not only ourselves but the entire universe.

9. Can Super-Consciousness explain phenomena like intuition or synchronicity?

Answer: Yes, Super-Consciousness can help explain phenomena like intuition and synchronicity. These experiences are often considered signs of a deeper connection between the individual and the universe. Intuition may arise when we tap into the collective knowledge of Super-Consciousness, accessing information beyond our immediate awareness. Synchronicities, or meaningful coincidences, can be seen as moments when the flow of Super-Consciousness aligns with our personal experiences, guiding us toward greater understanding or important life events.

10. How can understanding Super-Consciousness help me lead a more fulfilling life?

Answer: Understanding Super-Consciousness can help you lead a more fulfilling life by fostering a sense of connection and purpose. When you recognize that you are part of a greater whole, your actions and experiences take on new meaning. This perspective encourages you to live with greater mindfulness, compassion, and intention, knowing that your life is intertwined with the lives of others and the universe itself. By aligning with the flow of Super-Consciousness, you can create a life that is rich in meaning, connection, and spiritual fulfillment.

This chapter is designed to address some of the most common questions that arise when exploring the concepts of Super-Consciousness and spirituality. As you continue your journey, remember that the answers to these questions are not meant to be definitive but are intended to spark further exploration and reflection. The understanding of Super-Consciousness is an ongoing process, one that evolves with each new insight and experience.

Bibliography

The following sources have been referenced or inspired the ideas and concepts discussed in this book. They provide a foundation for the intersection of modern physics, spirituality, and philosophical thought, offering further reading and exploration for those interested in delving deeper into these topics.

Books

- **Capra, Fritjof.** *The Tao of Physics: An Exploration of the Parallels Between Modern Physics and Eastern Mysticism.* Shambhala Publications, 1975.

- **Hawking, Stephen.** *A Brief History of Time: From the Big Bang to Black Holes.* Bantam Books, 1988.

- **Penrose, Roger.** *The Emperor's New Mind: Concerning Computers, Minds, and the Laws of Physics.* Oxford University Press, 1989.

- **Schrödinger, Erwin.** *What Is Life? The Physical Aspect of the Living Cell.* Cambridge University Press, 1944.

- **Tegmark, Max.** *Our Mathematical Universe: My Quest for the Ultimate Nature of Reality.* Alfred A. Knopf, 2014.

Research Papers and Articles

- **Bell, John S.** "On the Einstein Podolsky Rosen Paradox." *Physics Physique Физика*, vol. 1, no. 3, 1964, pp. 195-200.

- **Bohm, David.** "A Suggested Interpretation of the Quantum Theory in Terms of 'Hidden' Variables I and II." *Physical Review*, vol. 85, no. 2, 1952, pp. 166-193.

- **Wheeler, John A., and Wojciech Hubert Zurek.** *Quantum Theory and Measurement.* Princeton University Press, 1983.

- **Aspect, Alain, Dalibard, Jean, and Roger, Gérard.** "Experimental Test of Bell's Inequalities Using Time-Varying Analyzers." *Physical Review Letters*, vol. 49, no. 25, 1982, pp. 1804-1807.

Websites and Online Resources

- **The Stanford Encyclopedia of Philosophy.** "Quantum Mechanics." plato.stanford.edu/entries/qm/ .

- **Nobel Prize in Physics.** "The Nobel Prize in Physics 2022." www.nobelprize.org/prizes/physics/2022/summary/.

- **Khan Academy.** "Quantum Mechanics." www.khanacademy.org/science/physics/quantum-physics .

Ancient Texts and Philosophical Works

- **The Bhagavad Gita.** Translated by Eknath Easwaran, Nilgiri Press, 2007.

- **Laozi.** *Tao Te Ching: A New English Version.* Translated by Stephen Mitchell, Harper & Row, 1988.

- **Buddha.** *The Dhammapada.* Translated by Thomas Byrom, Shambhala Publications, 1976.